INSTITUTION OF CIVIL ENGINEERS

MANAGEMENT DEVELOPMENT

IN THE

CONSTRUCTION INDUSTRY

Guidelines for the construction professional

Second edition

This edition has been updated and rewritten by:

Gavin Maxwell-Hart	Management Board, Institution of Civil Engineers Birse Construction Ltd
and supported by George Marsh	Management Board, Institution of Civil Engineers Galliford Try PLC

First edition, published February 1992

The Management Development Working Party for the first edition was:

C. Penny, Chairman,	Balfour Beatty Fairclough
J. R. Deacon	Cornwall county Council
K. C. Gower	Personnel Management Consultant to Laings Civil Engineering
D. Hattersley	R. O'Rourke & Sons Ltd
C. J. White	Scott Wilson Kirkpatrick, formerly with Travers Morgan

The authors have donated their fee for updating this work to RedR, the international charity working to relieve suffering in disasters.

Published for the Institution of Civil Engineers by Thomas Telford Publishing, Thomas Telford Ltd, 1 Heron Quay, London E14 4JD. URL: http://www.thomastelford.com

First published 1992
This edition first published 2001

A catalogue record for this book is available from the British Library

ISBN: 0 7277 2969 1

Printed and bound in Great Britain by Thanet Press, Margate

Foreword

The face of construction has been changed substantially during the past decade. This change has not yet stopped, in fact it is likely to accelerate for the foreseeable future but will result in an industry that will be modern, forward-thinking and progressive.

The profession is losing its arrogance and is at last realising that it is a 'service' industry and, subsequently, that its future success will depend on increasing professionalism, innovation and enhanced client understanding.

The industry has grasped the nettle in so far as there is a fast growing understanding that we must not only get to know our clients' requirements, but that we must then deliver to exceed their expectations. We also know that we must meet demanding technical, quality, time and cost parameters as agreed at time of purchase. We should perform in the way in which we expect other industries to perform. For example, if any of us invested in an expensive car but on the due date were told that the car was going to be three months late and that, unfortunately, we now had to pay an extra 20%, we would be exceedingly annoyed to say the least. If then, after late delivery, this new car had persistent quality and reliability problems we would demand compensation and retribution – why should the construction industry expect to be viewed any differently with regard to performance and delivery of the end product?

Our clients are becoming more sophisticated with ever increasing demands and requirements. We therefore cannot stand still, we must constantly seek development and refinements to the way in which we perform, operate and deliver. This means that we have no option other but to continue to increase our agility, our management expertise and our technical innovation and creativity.

Enhanced management and control will go a long way towards helping us to thrive in this future world. Effective training and development is a prerequisite to create the vision, abilities and skills to match future demands.

The education system gives excellent grounding and high standards of knowledge acquisition and the average UK construction novice is therefore basically well equipped. He or she must be guided to develop the right skills to manage our business in a way that maximises our future delivery and returns. This requires a bolder and more committed attitude to the training and development of our members to enhance the interface with our clients, our management and control skills and our professional expertise.

Better training, better skills and better management will help us to control our own destiny more closely.

Contents

Preface

This book was first written in 1991 and has now been brought up to date after a period of some ten years.

The speed at which both the industry has developed and the role of the construction professional has advanced within this decade has been rapid.

The actual degree of management development became overwhelmingly apparent during this rewrite and is well demonstrated by the increased responsibilities and levels of competence that we must now all shoulder.

- There were originally ten key roles, now there are twelve – an increase of some 20%, probably not that surprising, but
- the number of elements, within these key roles, within which it is necessary to gain a high competence level, was 75, now it is 143 – *a staggering increase of 90%!*

Virtually all subject matters within this increase have now become fundamental aspects through which we all carry out our daily working lives, our business management and ability and skill development. A few of these are given below:

- CDM and construction safety regulations
- client focus and understanding
- partnering or alliancing
- PFI, PPP or other forms of creative financing
- risk identification, evaluation and management
- benchmarking, performance measurement and continuous improvement
- innovation and value management
- supply chain management
- zero defects
- environmental sustainability.

Therefore, it is not surprising that we must gain greater understanding of substantially more issues and the part that they all play within in the industry. Probably more importantly, this in turn leads to the requirement that we gain greater expertise in the management and control of people, systems and processes in order to ensure successful results.

Thus, management development has become a far more important part of the construction professional's duties and career development. It is essential, not only for his or her individual well-being but also that of the construction profession as a whole.

1. The construction profession of the future

The nature of the change that the construction industry has undergone in the 1990s has been to bring us up to date with the advancements of the 'outside' world and other more progressive industries and sectors. In simple terms, we had lost our way and in order to survive against external opposition, we were forced to play 'catch-up'.

The pace and format of change in the next decade will be driven from within the industry, to re-establish ourselves as a leading profession exhibiting its characteristics of flair, dynamism and progression. We will not be waiting for 'others' to tell us how to operate, but will be dictating our own pace and our own requirements.

As UK construction professionals we still have skills and abilities that allow us to progress to the highest levels of achievement. We have excellent grounding in education and training and, once the constraints are removed, we instinctively have the capacity to excel!

We will meet the challenge of transforming our industry.

So where is our future?

This is difficult to predict with any accuracy but, without a doubt, we will all need to develop a much stronger grasp of computers, the Internet, e-commerce, standardisation, prefabrication, automation and robotics. We will manage remotely and need to control a wider range of integrated projects through far more complex interfaces and risk and financial solutions. We will need to manage substantially more integrated relationships with our clients in order to understand and forward deliver to meet their clients' future needs.

We will need to have transformed the way in which we think, the way in which we behave and the way in which we are structured. The industry will be 'de-skilled' and then 're-skilled' to change its skills base to one founded on highly skilled professionals who manage the systems and processes that bring together and control the various components and outsourced supply chain.

Greater and extended management and management skills for all our members will need to come to the fore. This will demand greater efficiencies and skills from our members, but we will train, develop and equip them properly to continue to advance the industry.

2. Introduction

Management development needs differ from person to person, depending on their particular circumstances and aspirations. It is therefore important that individuals take responsibility in the first instance for their own development.

The key roles that represent the fundamentals of modern construction 'management' have been deemed to fall into twelve categories, each with a number of elements to reflect the actual responsibilities in which we need to have expertise. These elements require different levels of competence at different career stages and model competencies have been produced that are representative across all sectors of the industry.

The key roles and the associated model competencies are central to these guidelines. They provide a diagnostic tool that invites self-assessment and comparison, allowing individuals to determine for themselves their management development needs.

It must be stressed that the model competencies are not in any way prescriptive. They serve as a means by which individuals can test their own development and identify further areas in which they could usefully apply themselves.

During the preparation of the original document it became apparent that there was a commonality in management development within all sectors of the industry – this has been reinforced during the updating of the document. Further, it has been found that the young managers of today are prepared to move more readily across sectors in order to develop their careers. For this reason, sector-specific elements have been omitted from the key roles, although organisations are encouraged to introduce such additional elements to meet their specific requirements.

3. Key roles

The twelve key roles of management in the construction industry and the key purpose for each are as follows:

Key role 1: **Corporate management**
Key purpose: Creation of long-term corporate vision and objectives, development and progression of management organisational structure and understanding of the implications of external factors that will affect the future strength and direction of an organisation.

Key role 2: **Business management**
Key purpose: The practical management of an organisation necessary to achieve corporate objectives and targets through competitive advantage and the creation of processes and systems to benchmark, learn and ensure continuous improved performance and betterment.

Key role 3: **Financial and management systems**
Key purpose: The systems necessary to monitor and control all aspects of the financial well-being of an organisation with the view to maximising profit.

Key role 4: **Promotion and business development**
Key purpose: The marketing and promotional management to match business strengths and objectives with the most appropriate market sectors and clients to maximise business opportunities and potential.

Key role 5: **Communications and presentations**
Key purpose: Effective communication both verbal and written with business associates, clients, the media and the public.

Key role 6: **The client and relationships**
Key purpose: The understanding to build client relationships by trust, quality and commitment in order to be able to deliver projects or services to exceed clients' needs and requirements, thus allowing all parties to enhance their own profit and wealth creation.

Key role 7: **Respect for people**
Key purpose: The skills necessary to create a sustainable and viable people culture and to select, motivate and lead people in order to maximise their performance for their organisation and for themselves

Key role 8: **Project management**
Key purpose: The multi-faceted responsibilities necessary to programme, monitor and control all aspects of a project from conception to successful handover in line with clients' requirements, one's own organisational objectives and the implementation of best practice.

Key role 9: **Professional, commercial and contractual practice**

Key purpose: Development of the professional, commercial, contractual and legal factors affecting the main undertakings of an organisation.

Key role 10: **Information and communication technology**

Key purpose: Personal skill development coupled with appreciation and management to enable the use of computer technology to enhance business effectiveness through better communication, engineering, management, control and procurement systems.

Key role 11: **Health, safety and welfare, quality and environment**

Key purpose: The knowledge and understanding of all responsibilities and accountabilities, both legal and practical, and the use of all necessary procedures, systems and training to ensure that the highest appropriate standards are achieved.

Key role 12: **The construction profession in society**

Key purpose: The responsibilities of all construction professionals to maximise our service to all people in society, and thereby promote the industry.

In ordering the key roles no attempt has been made to ascribe relative priorities. Some deal with interpersonal skills that are important to managers, others deal with techniques or instruments of management.

The elements of each key role have been selected as both important and relevant. They have been limited to those identified as applicable to all sectors of the industry. The number assigned to elements should not be construed as implying levels of significance. It is important to establish clear interpretations of the elements and therefore they are defined for the purposes of this document in the Glossary to obviate the likelihood of them meaning different things to different people.

4. Management levels

The level of competence in the elements is dependent on particular career stages, and so these have been redefined to match present-day management structures and practices.

The career stages have been adopted with respect to job titles that have universal usage in the industry and also indicate typical levels of responsibility so that, irrespective of sector, individuals can identify their present level and their next level.

The job titles and responsibility used are as follows:

Executive management

Typical job titles:	Chief executive, senior director, senior partner or most senior professional leader in an organisation or deputy in a large organisation
Typical responsibilities:	Strategic management of organisations, in addition to the overall management to achieve the business plan objectives for the current year

Senior management

Typical job titles:	Divisional director, partner or associate, chief engineer, chief technical or professional director
Typical responsibilities:	Management of a major sector or function of an organisation; implementing the strategic and short-term objectives with a reasonable level of autonomy and control to achieve the annual targets and aims Also responsible for the overall training and development of the management team to ensure continuous performance improvement

Middle management

Typical job titles:	Divisional manager, contracts manager, senior project manager, senior engineering or function manager, senior technical or professional manager
Typical responsibilities:	Management of a major project or a number of concurrent projects or assignments, including responsibility for the teams working on them, to achieve desired targets Also responsible for implementing training and development and the incorporation of 'best practise'

Supervisory management

Typical job titles:	Project manager/agent, construction manager, function manager, project or assignment engineer or professional person
Typical responsibilities:	Management of a small project or a small team, requiring professional and financial judgement and needing minimum supervision

Junior management

Typical job titles:	Junior and graduate engineer or professional person

Typical responsibilities:	Obtaining experience and professional training at the discretion of others, normally in the early stages of their career, taking key roles in teams as individuals, exercising direction of others on a project or part of a large project

5. Levels of competence

For continuity the four levels of competence have been maintained, although to enhance individual progression, they have been slightly redefined. Each element of the key roles has the appropriate level ascribed at progressive stages of career development.

The levels of competence are:

A	Appreciation	Know what is meant by a term and what its purpose and objective is.
K	Knowledge	Understand in detail the principles of the topic and how and where they are applied
E	Experience	Have acquired working knowledge and skill
B	Ability	Be able to fully apply skill with desired results

The indication of competence at varying career stages for each element in all the key roles produces the model competencies given in section 7.

The model competencies accept that competence is either being obtained, kept current or being lost. These conditions are illustrated by arrows to represent progressive acquisition of competence, the level at which competence is to be sustained and that the competence is likely to regress from the initial peak and/or previous level indicated.

6. How to use this document

Step 1: Current management level

Identify your current management level from section 4. In most cases either the job title you have will match one of those in section 4 or you will be able to match your responsibilities with those shown as typical. However, if you consider that you fall between the categories listed or are not sure whether you are on one level or the next, opt for the higher level.

Step 2: Self-assessment

Self-assessment and audit forms are given at the back of this document for you to conduct your self-assessment. There is a form for each key role and you can use each form three times. These forms allow you to identify your needs.

Using A, K, E or B as defined in section 5, consider your competence in each element in all the key roles. You do not have to assess against key roles or elements that you know have absolutely no relevance to you or the job you do. You need to be aware that they do have relevance to 'management' and may therefore become relevant in time.

An example of self-assessment is shown in Fig. 1.

Step 3: Comparison with model competencies

Overlay your self-assessment and audit form on the appropriate model competency in section 7 and determine whether you match, exceed or fall short of the competency given for the level at which you are assessing yourself.

Fig. 1. Example of self-assessment

KEY ROLE 1	CORPORATE MANAGEMENT																			
	ELEMENTS	CORPORATE STRATEGY	BUSINESS PLANS AND OBJECTIVES	ORGANISATION AND MANAGEMENT STRUCTURE	ROLES OF OFFICERS	SPECIALIST FUNCTIONS AND THEIR MANAGEMENT	MEMORANDUM AND ARTICLES	MERGERS AND ACQUISITIONS	FINANCIAL ARRANGEMENTS	BONDS AND WARRANTY FACILITIES	POLITICAL AND EXTERNAL INFLUENCES	STOCK MARKET AND CITY	VALUES AND CULTURE							
MANAGEMENT LEVEL		1	2	3	4	5	6	7	8	9	10	11	12	13	14	15	16	17	18	19
Management level at which assessment carried out:	B		B																	
Middle Management Date: 9. 1. 01	E			E						E			E							
	K				K	K	K		K		K									
	A	A						A				A								

KEY ROLE 1		CORPORATE MANAGEMENT																		
MODEL COMPETENCIES	ELEMENTS	CORPORATE STRATEGY	BUSINESS PLANS AND OBJECTIVES	ORGANISATION AND MANAGEMENT STRUCTURE	ROLES OF OFFICERS	SPECIALIST FUNCTIONS AND THEIR MANAGEMENT	MEMORANDUM AND ARTICLES	MERGERS AND ACQUISITIONS	FINANCIAL ARRANGEMENTS	BONDS AND WARRANTY FACILITIES	POLITICAL AND EXTERNAL INFLUENCES	STOCK MARKET AND CITY	VALUES AND CULTURE							
MANAGEMENT LEVEL		1	2	3	4	5	6	7	8	9	10	11	12	13	14	15	16	17	18	19
EXECUTIVE MANAGEMENT	B	B	B	B	B	B	B	B	B	B	B	B	B							
- Chief Executive - Senior Director	E																			
- Senior Partner	K																			
- Senior Professional Person	A																			
SENIOR MANAGEMENT	B		B	B	B								B							
- Divisional Director - Partner/Associate	E	E				E	E	E	E	E	E	E								
- Chief Engineer - Chief Technical/Professional	K																			
Person	A																			
MIDDLE MANAGEMENT	B																			
- Divisional Manager Senior Project Manager	E		E	E	E								E							
- Senior Engineer - Senior Technical/Professional	K	K				K	K	K	K	K	K	K								
Person	A																			

INITIAL PEAK COMPETENCE LEVEL

		1	2	3	4	5	6	7	8	9	10	11	12	13	14	15	16	17	18	19
Management level at which Assessment carried out:	B		B																	
Middle Management	E			E						E			E							
Date: 9. 1. 01	K				K	K	K		K		K									
	A	A						A				A								
Assessment as model				X		X	X		X		X		X							
Assessment greater than model			X							X										
Assessment less than model		X			X			X				X								
ELEMENTS FOR ACTION		*			*			*				*								
Management level at which Assessment carried out:	B																			
	E																			
Date:	K																			
	A																			
Assessment as model																				
Assessment greater than model																				
Assessment less than model																				
ELEMENTS FOR ACTION																				

*** Elements for action**

Fig. 2. Example of audit

Record your assessment on the audit part of the self-assessment and audit forms, as shown in the example Fig. 2. By folding the forms along the dotted lines you can make direct comparison of your competencies with those shown in the model and record the result.

An example of an audit is shown in Fig. 2.

Step 4: Identify priorities

From your audit, identify elements that you consider require action. These may not be just those for which your assessment is less than the model competence: for example, your assessment may well match the model for a certain element that you regard as very important and therefore requiring further development. No matter how much you try to limit the number of elements you identify as requiring action, it is likely that there will be too many for you to deal with them all in a reasonable period of time. To reduce the number to something manageable you will need to adopt a procedure that allows you to identify your priorities.

The real test of priorities is derived from your current responsibilities and whether or not you need the particular competence you are considering to do your job better, more efficiently or effectively, or in preparation for your next career step.

Before your final decision, it is important also that you debate your chosen priorities with your line manager/HR manager during your annual staff performance appraisal. This will ensure that you match your expectations and requirements with those of your employer. (These appraisals can also be used at a later stage, to check your Action Plan and audit the progress that you are actually achieving.)

To help reach such conclusions the answers which you give to the following three questions will help:

- is the competency relevant to your job?
- do you need the competency now or later?
- how much knowledge, experience and ability do you already possess?

If you conclude you would be more effective in your job were you skilled in a particular element, that you need that skill now and that you do not have sufficient knowledge, experience or ability in the subject, it must become a priority for action.

Alternatively, you may decide that there is a matter about which your self-assessment fell short of the model and that it is something that you can see would be useful. However, you conclude that it has no immediate relevance in terms of your current responsibilities. In such circumstances you cannot give it priority.

Your aim is to arrive at the minimum number of priority areas for action. You will be wise to address no more than three items at a time. You can bring fresh items into consideration when you have achieved those you have already addressed.

Step 5: Taking action

Having determined areas for development, you must establish an action plan that sets a timetable, addresses how you will achieve your objectives and tests the appropriateness of your decisions. This is in many ways the most important step. Section 8 provides guidance on how to develop an action plan.

If your self-assessment and audit produces few or no elements requiring action at the management level at which you made the assessment, you should repeat the process for the next career stage and identify elements for action that are of a developmental nature.

Summary

Using this document involves a five-step procedure.

- Determine current or next level of management.
- Carry out a self-assessment against the key roles.
- Compare assessment against model competencies.
- Audit results and determine priorities.
- Take action.

You can use this procedure for basic self-assessment, or whenever you feel your circumstances have changed for any reason, i.e. you have increased responsibilities, a change of employer or have achieved previous actions which you set yourself.

Although this document is not intended to be in any way mandatory or prescriptive, it provides candidates preparing for admission to membership of their various professional bodies with a clear insight of the order of management skills required for Professional Review. It would be sensible, therefore, for potential professional candidates to enclose a copy of a self-assessment, audit and action plan as part of their submission. Candidates would normally be expected at their Professional Review to have the competencies equal to or greater than those indicated for junior management

7. Model competencies

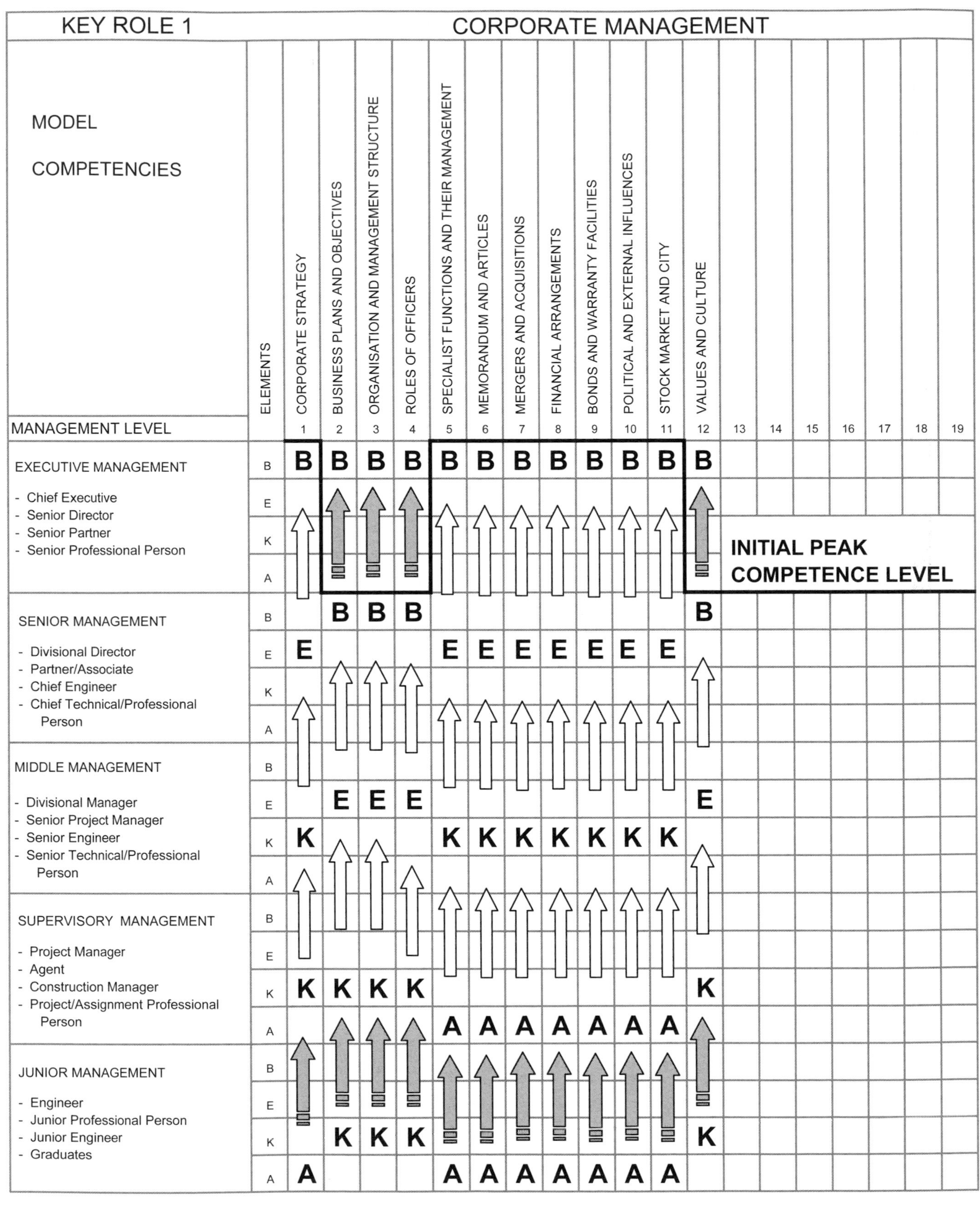

MANAGEMENT DEVELOPMENT

KEY ROLE 1 — CORPORATE MANAGEMENT

MODEL COMPETENCIES

MANAGEMENT LEVEL	ELEMENTS	1 CORPORATE STRATEGY	2 BUSINESS PLANS AND OBJECTIVES	3 ORGANISATION AND MANAGEMENT STRUCTURE	4 ROLES OF OFFICERS	5 SPECIALIST FUNCTIONS AND THEIR MANAGEMENT	6 MEMORANDUM AND ARTICLES	7 MERGERS AND ACQUISITIONS	8 FINANCIAL ARRANGEMENTS	9 BONDS AND WARRANTY FACILITIES	10 POLITICAL AND EXTERNAL INFLUENCES	11 STOCK MARKET AND CITY	12 VALUES AND CULTURE	13	14	15	16	17	18	19
EXECUTIVE MANAGEMENT - Chief Executive - Senior Director - Senior Partner - Senior Professional Person	B	B	B	B	B	B	B	B	B	B	B	B	B							
	E																			
	K													INITIAL PEAK COMPETENCE LEVEL						
	A																			
SENIOR MANAGEMENT - Divisional Director - Partner/Associate - Chief Engineer - Chief Technical/Professional Person	B		B	B	B								B							
	E	E				E	E	E	E	E	E	E								
	K																			
	A																			
MIDDLE MANAGEMENT - Divisional Manager - Senior Project Manager - Senior Engineer - Senior Technical/Professional Person	B																			
	E		E	E	E								E							
	K	K				K	K	K	K	K	K	K								
	A																			
SUPERVISORY MANAGEMENT - Project Manager - Agent - Construction Manager - Project/Assignment Professional Person	B																			
	E																			
	K	K	K	K	K								K							
	A					A	A	A	A	A	A	A								
JUNIOR MANAGEMENT - Engineer - Junior Professional Person - Junior Engineer - Graduates	B																			
	E																			
	K		K	K	K								K							
	A	A				A	A	A	A	A	A	A								

LEVELS OF COMPETENCE

A Appreciation: Know what is meant by the term and what its purpose is.

K Knowledge: Understand in some detail the principles of the topic and how they are applied.

E Experience: Acquisition of knowledge and skill.

B Ability: Application of skill with satisfactory results.

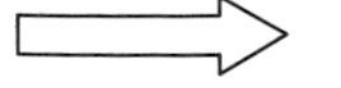 PROGRESSIVE  SUSTAINED  REGRESSIVE

MANAGEMENT DEVELOPMENT

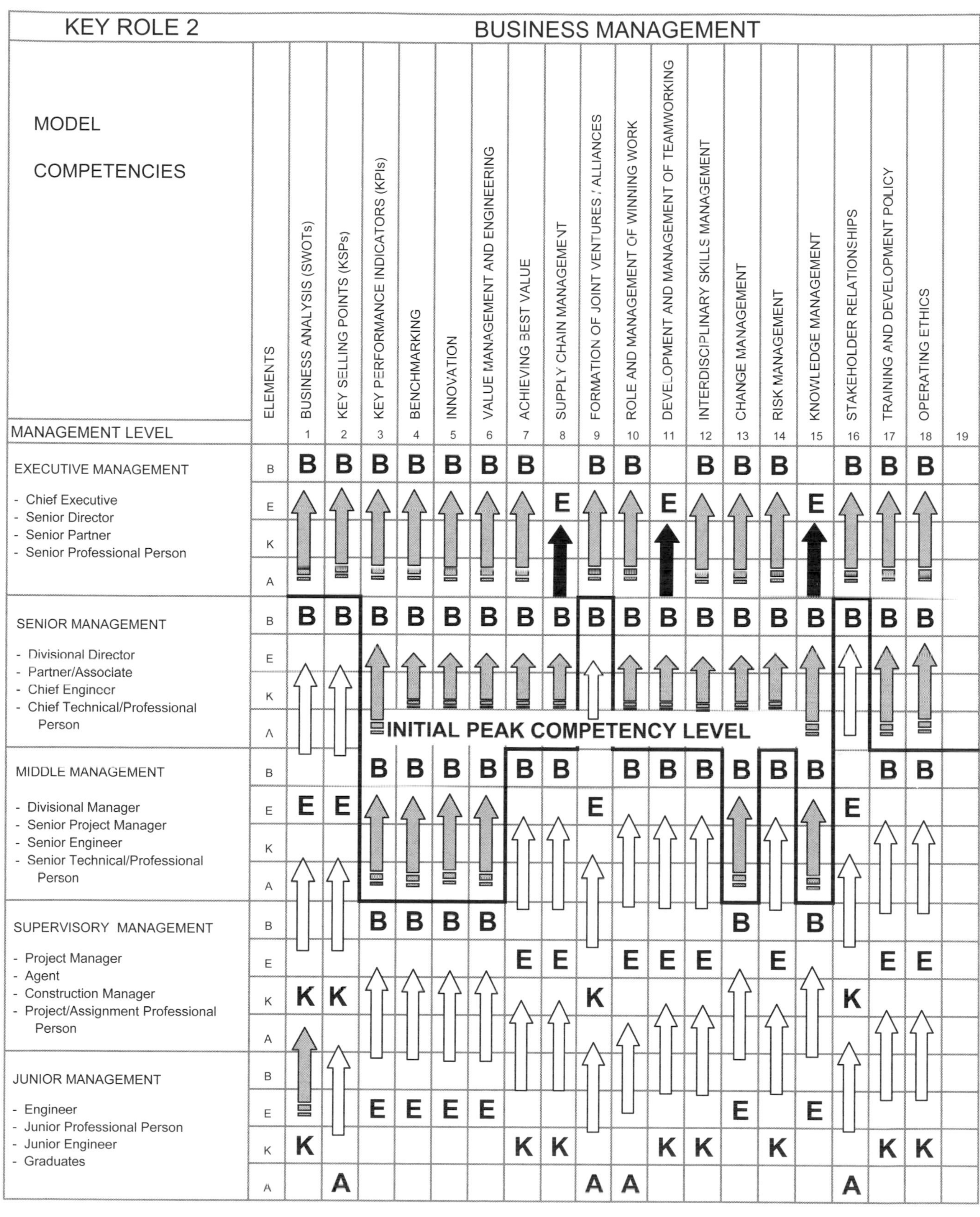

LEVELS OF COMPETENCE

A Appreciation: Know what is meant by the term and what its purpose is.

K Knowledge: Understand in some detail the principles of the topic and how they are applied.

E Experience: Acquisition of knowledge and skill.

B Ability: Application of skill with satisfactory results.

 PROGRESSIVE

 SUSTAINED

 REGRESSIVE

MANAGEMENT DEVELOPMENT

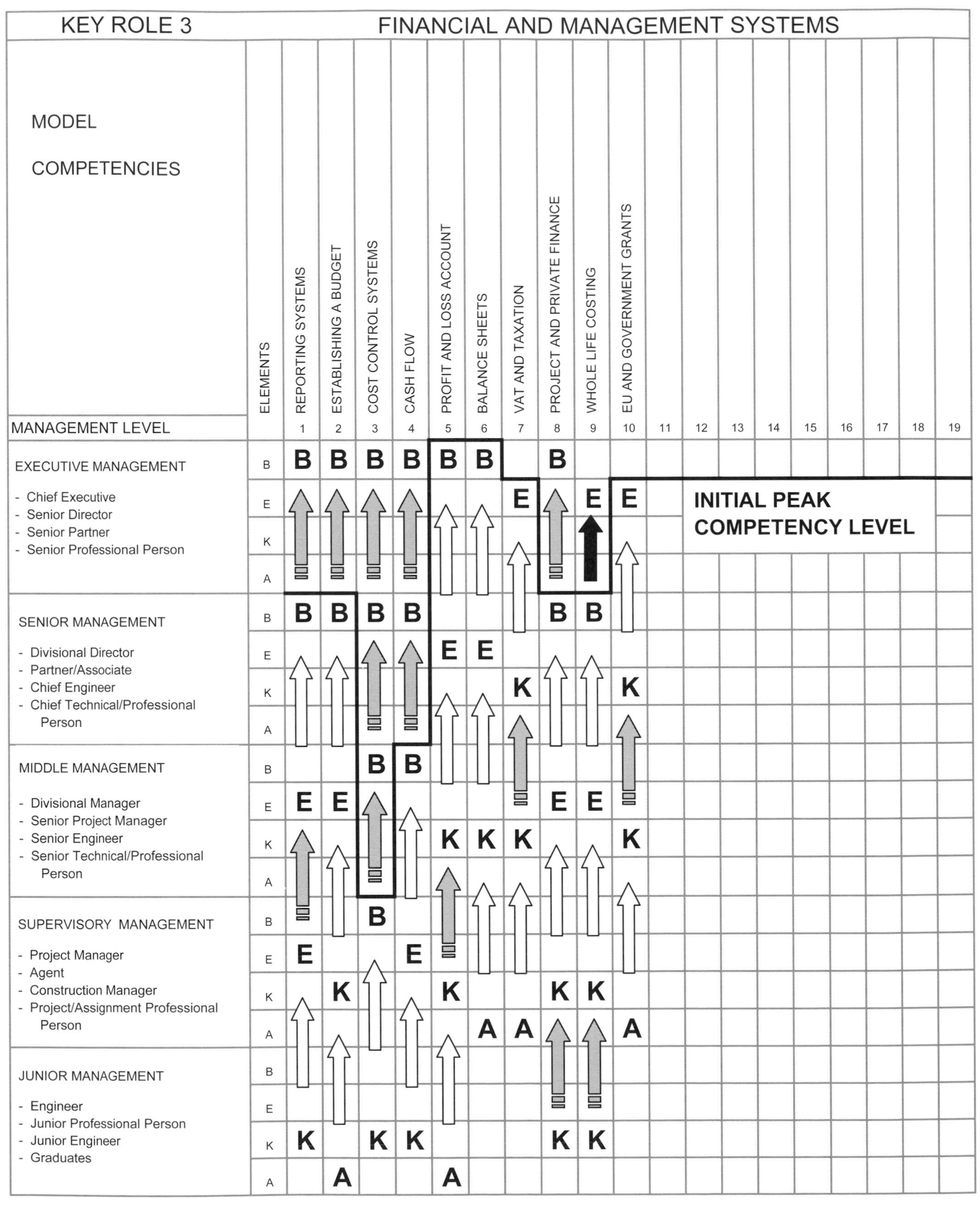

LEVELS OF COMPETENCE

A Appreciation: Know what is meant by the term and what its purpose is.

K Knowledge: Understand in some detail the principles of the topic and how they are applied.

E Experience: Acquisition of knowledge and skill.

B Ability: Application of skill with satisfactory results.

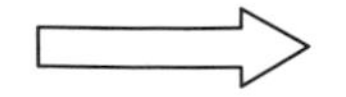 PROGRESSIVE

 SUSTAINED

 REGRESSIVE

MANAGEMENT DEVELOPMENT

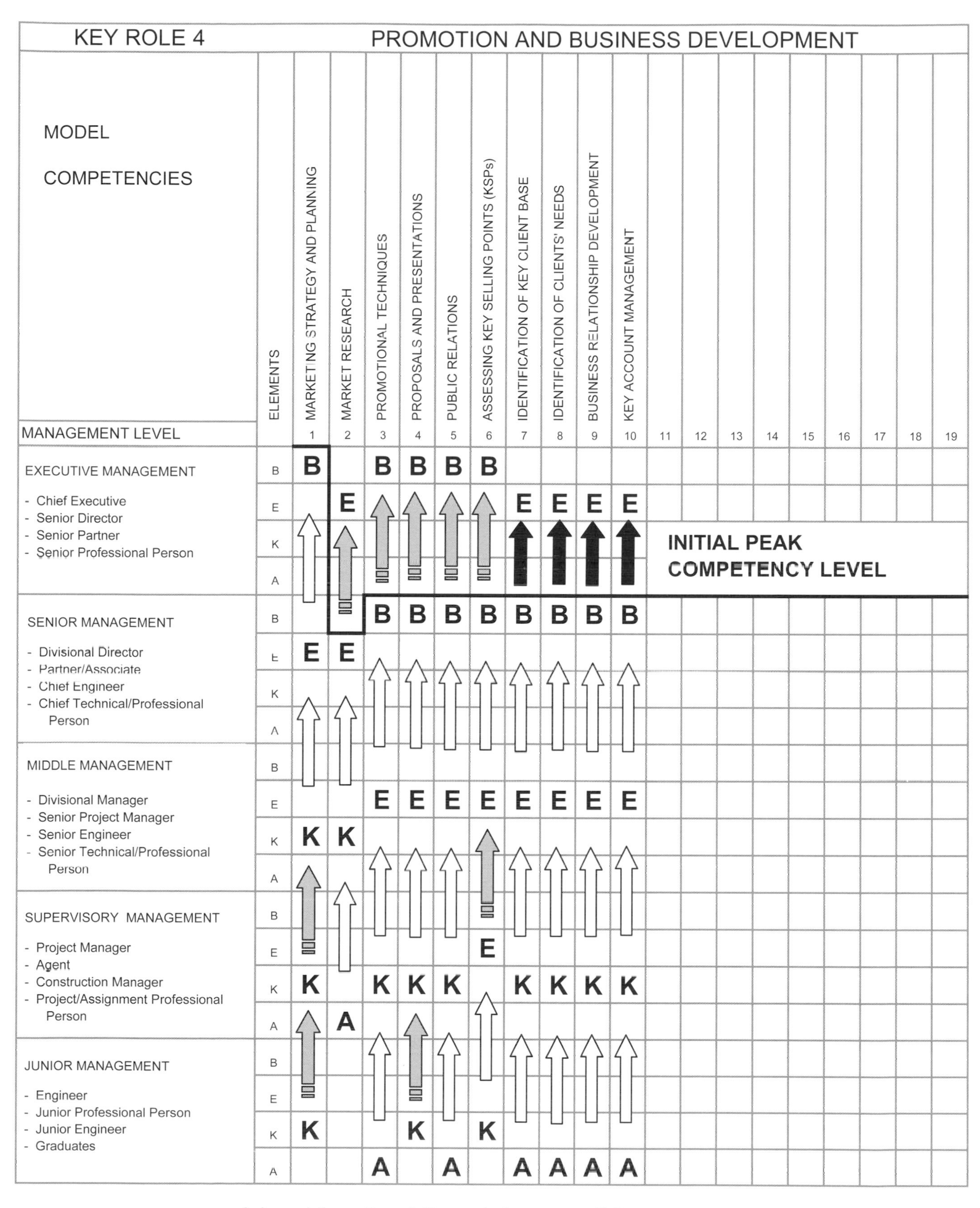

LEVELS OF COMPETENCE

A Appreciation: Know what is meant by the term and what its purpose is.

K Knowledge: Understand in some detail the principles of the topic and how they are applied.

E Experience: Acquisition of knowledge and skill.

B Ability: Application of skill with satisfactory results.

 PROGRESSIVE  SUSTAINED  REGRESSIVE

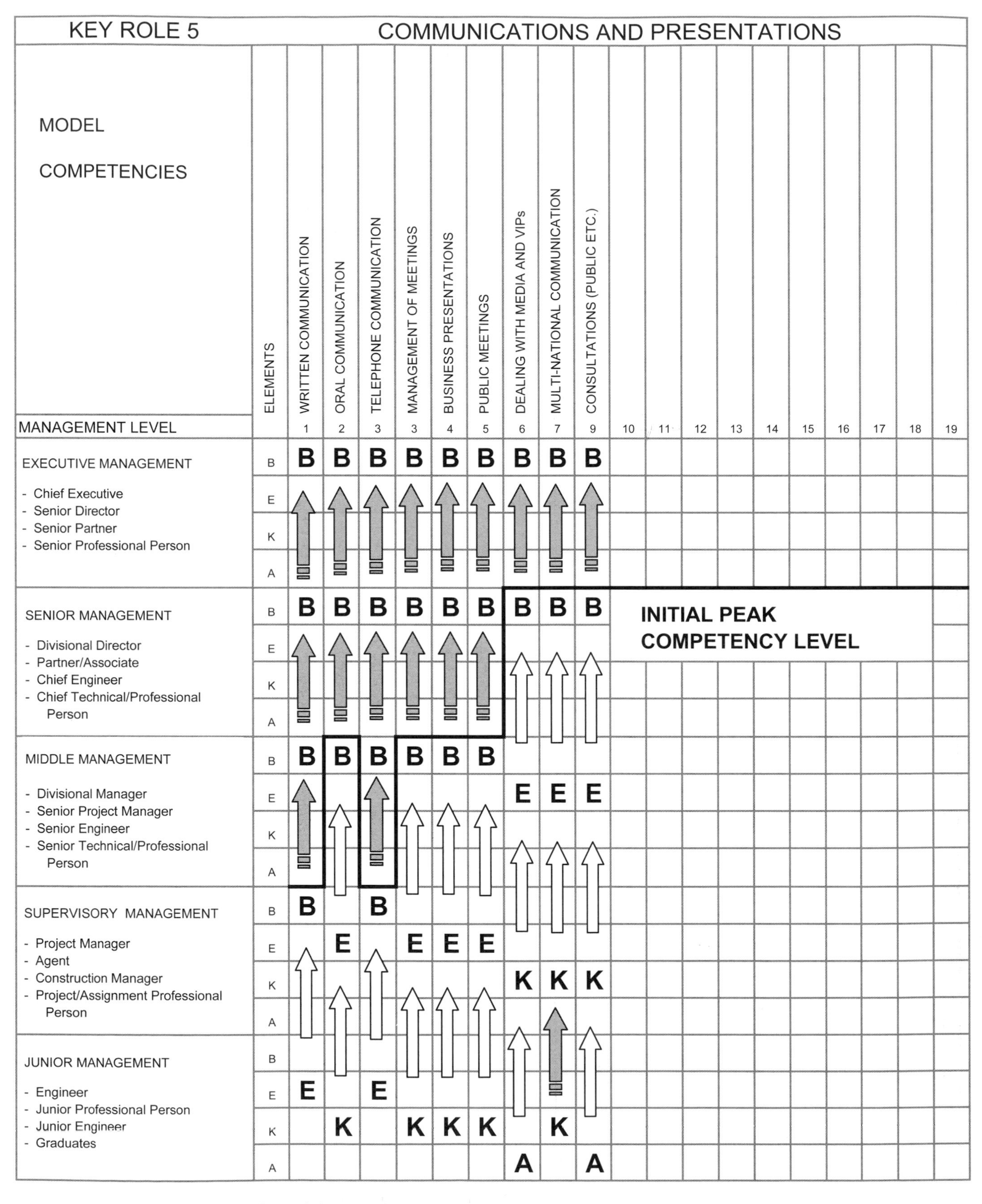

LEVELS OF COMPETENCE

A Appreciation: Know what is meant by the term and what its purpose is.

K Knowledge: Understand in some detail the principles of the topic and how they are applied.

E Experience: Acquisition of knowledge and skill.

B Ability: Application of skill with satisfactory results.

 PROGRESSIVE

 SUSTAINED

 REGRESSIVE

MANAGEMENT DEVELOPMENT

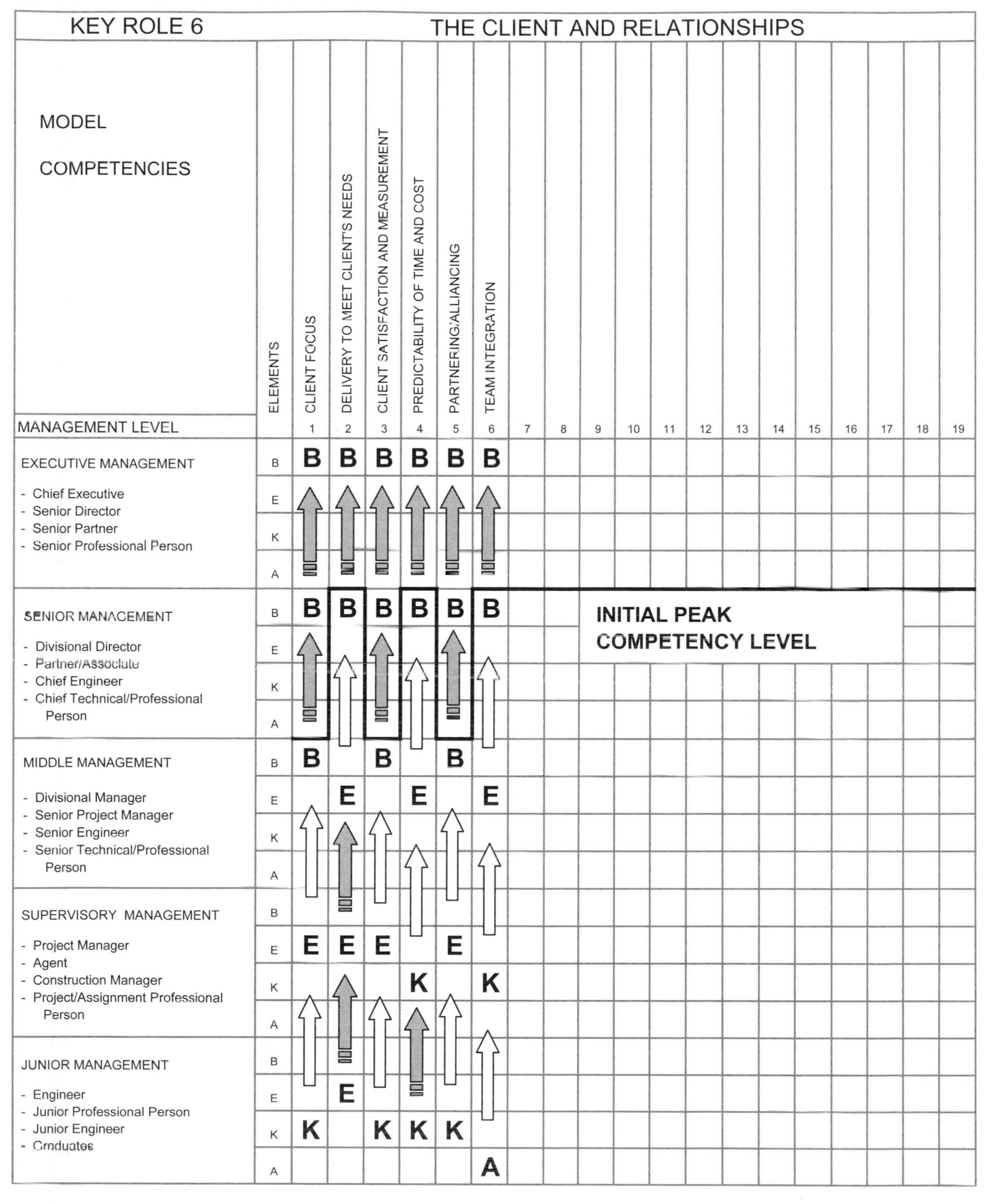

LEVELS OF COMPETENCE

A Appreciation: Know what is meant by the term and what its purpose is.

K Knowledge: Understand in some detail the principles of the topic and how they are applied.

E Experience: Acquisition of knowledge and skill.

B Ability: Application of skill with satisfactory results.

 PROGRESSIVE

 SUSTAINED

 REGRESSIVE

MANAGEMENT DEVELOPMENT

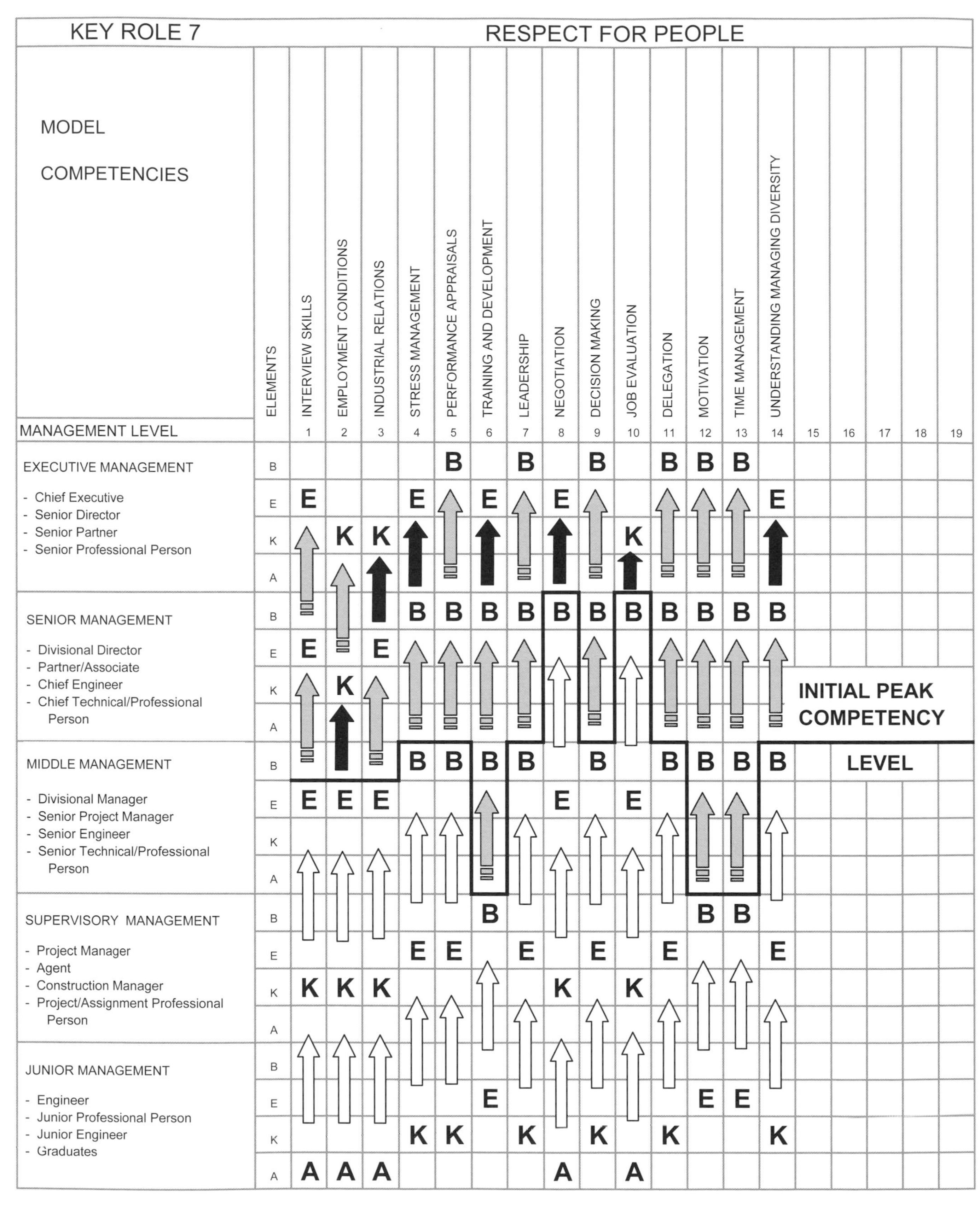

LEVELS OF COMPETENCE

A Appreciation: Know what is meant by the term and what its purpose is.

K Knowledge: Understand in some detail the principles of the topic and how they are applied.

E Experience: Acquisition of knowledge and skill.

B Ability: Application of skill with satisfactory results.

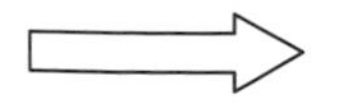 PROGRESSIVE

 SUSTAINED

 REGRESSIVE

MANAGEMENT DEVELOPMENT

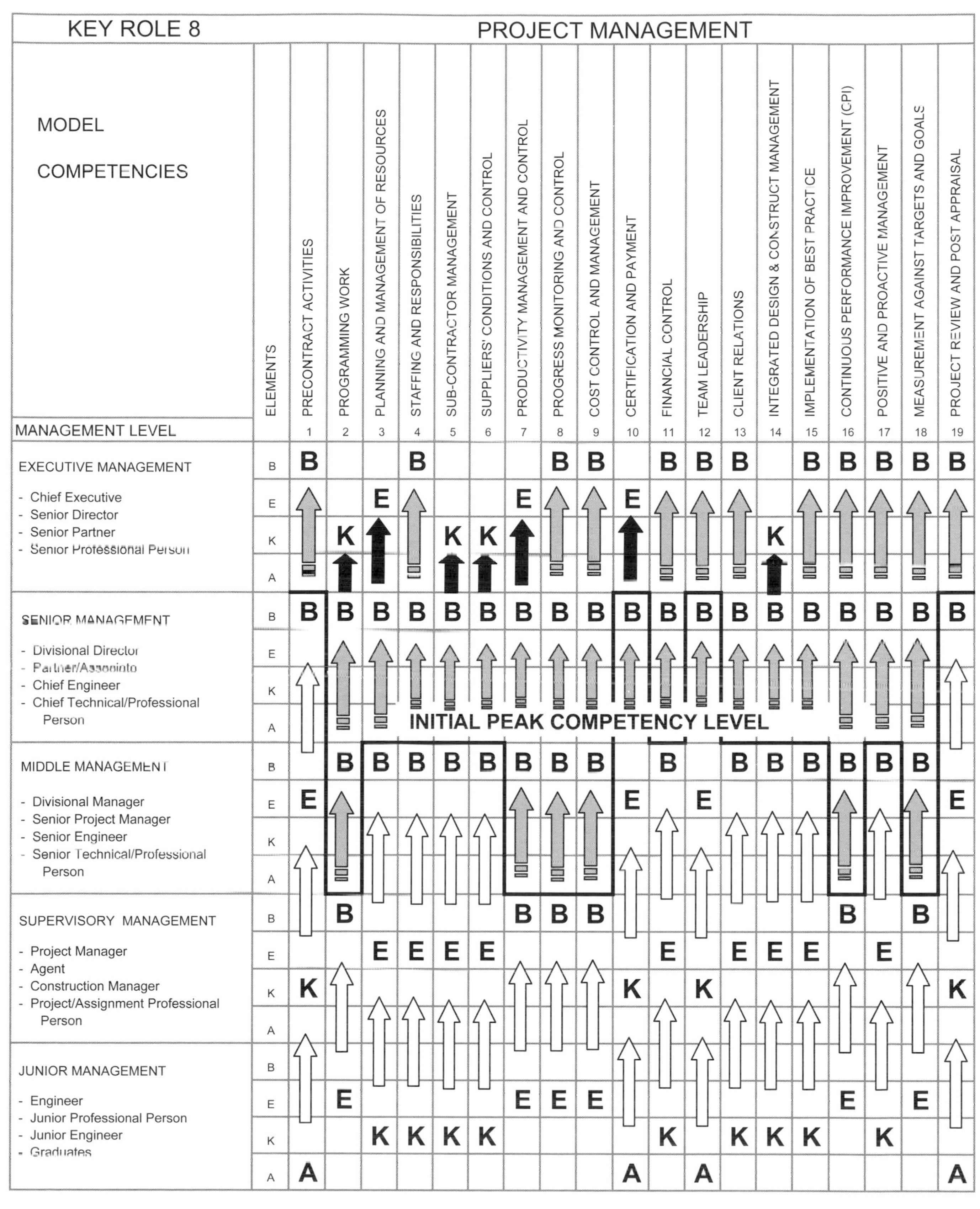

LEVELS OF COMPETENCE

A Appreciation: Know what is meant by the term and what its purpose is.

K Knowledge: Understand in some detail the principles of the topic and how they are applied.

E Experience: Acquisition of knowledge and skill.

B Ability: Application of skill with satisfactory results.

 PROGRESSIVE

 SUSTAINED

 REGRESSIVE

MANAGEMENT DEVELOPMENT

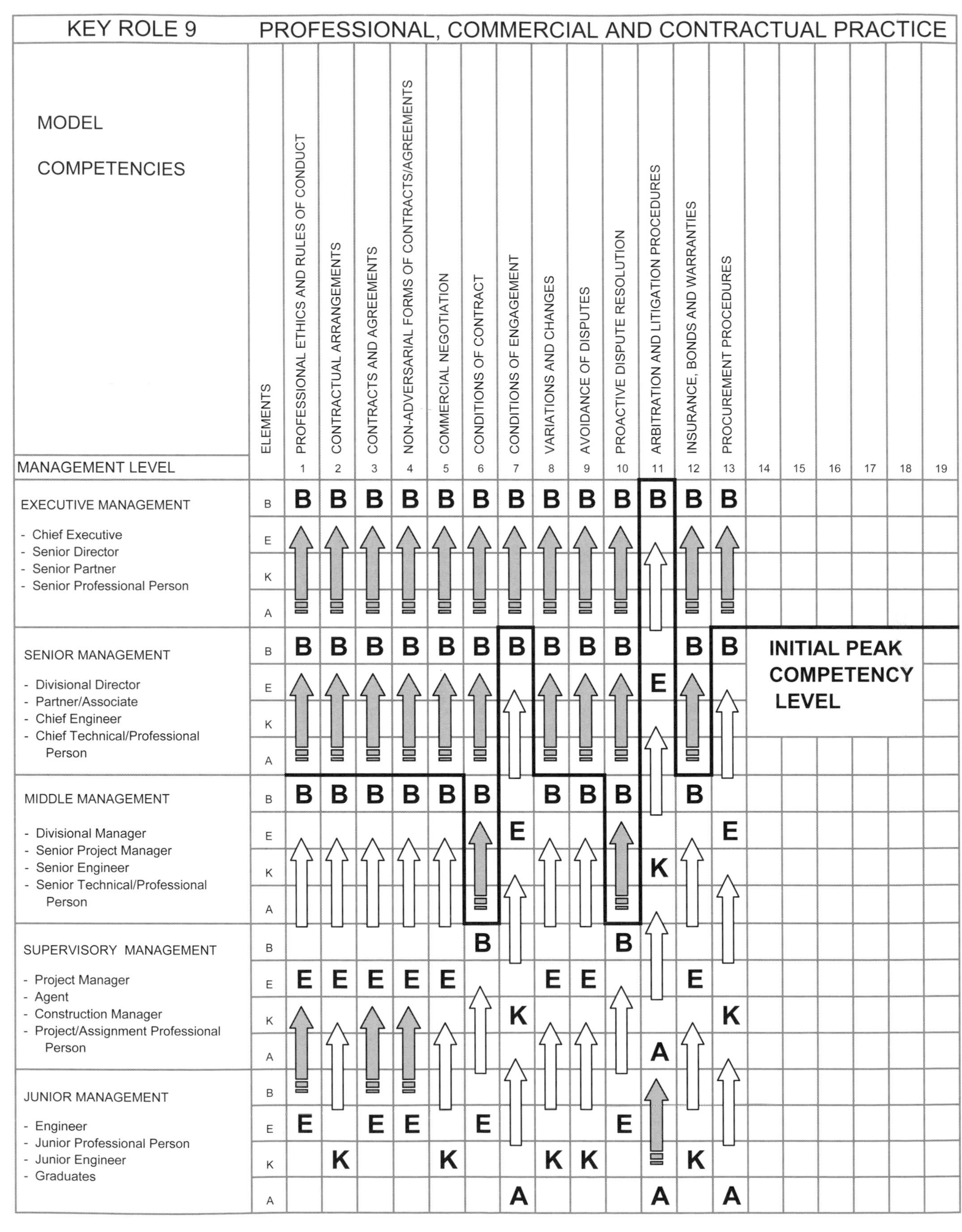

LEVELS OF COMPETENCE

A Appreciation: Know what is meant by the term and what its purpose is.

K Knowledge: Understand in some detail the principles of the topic and how they are applied.

E Experience: Acquisition of knowledge and skill.

B Ability: Application of skill with satisfactory results.

 PROGRESSIVE  SUSTAINED  REGRESSIVE

MANAGEMENT DEVELOPMENT

KEY ROLE 10 — INFORMATION AND COMMUNICATION TECHNOLOGY

MODEL COMPETENCIES / MANAGEMENT LEVEL	ELEMENTS	**PERSONAL COMPETENCE**	1 PERSONAL ICT SKILLS	2 SPECIALISED SOFTWARE APPLICABLE TO ROLE	3 ELECTRONIC COMMUNICATIONS	4	5	6	7	8	**MANAGEMENT AND APPRECIATION**	9 USE AND MANAGEMENT OF ICT RESOURCES	10 USE AND MANAGEMENT OF E-COMMERCE	11 USE AND MANAGEMENT OF INTERNET APPLICATIONS	12 DEVELOPMENT OF IT VISION	13 MANAGEMENT OF OUTSOURCED ICT SERVICES	14	15	16	17
EXECUTIVE MANAGEMENT - Chief Executive - Senior Director - Senior Partner - Senior Professional Person	B																			
	E		E		E							E	E	E	E	E				
	K			K																
	A																			
SENIOR MANAGEMENT - Divisional Director - Partner/Associate - Chief Engineer - Chief Technical/Professional Person	B				B							B	B	B	B	B				
	E		E	E																
	K																			
	A																			
MIDDLE MANAGEMENT - Divisional Manager - Senior Project Manager - Senior Engineer - Senior Technical/Professional Person	B				B															
	E		E	E								E	E	E	E	E				
	K																			
	A																			
SUPERVISORY MANAGEMENT - Project Manager - Agent - Construction Manager - Project/Assignment Professional Person	B																			
	E		E		E															
	K			K								K	K	K	K	K				
	A																			
JUNIOR MANAGEMENT - Engineer - Junior Professional Person - Junior Engineer - Graduates	B																			
	E		E		E															
	K																			
	A			A								A	A	A	A	A				

INITIAL PEAK COMPETENCY LEVEL

LEVELS OF COMPETENCE

A Appreciation: Know what is meant by the term and what its purpose is.

K Knowledge: Understand in some detail the principles of the topic and how they are applied.

E Experience: Acquisition of knowledge and skill.

B Ability: Application of skill with satisfactory results.

 PROGRESSIVE

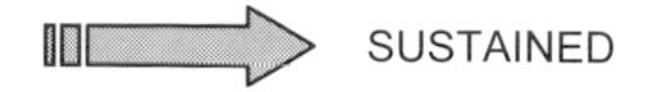 SUSTAINED

 REGRESSIVE

MANAGEMENT DEVELOPMENT

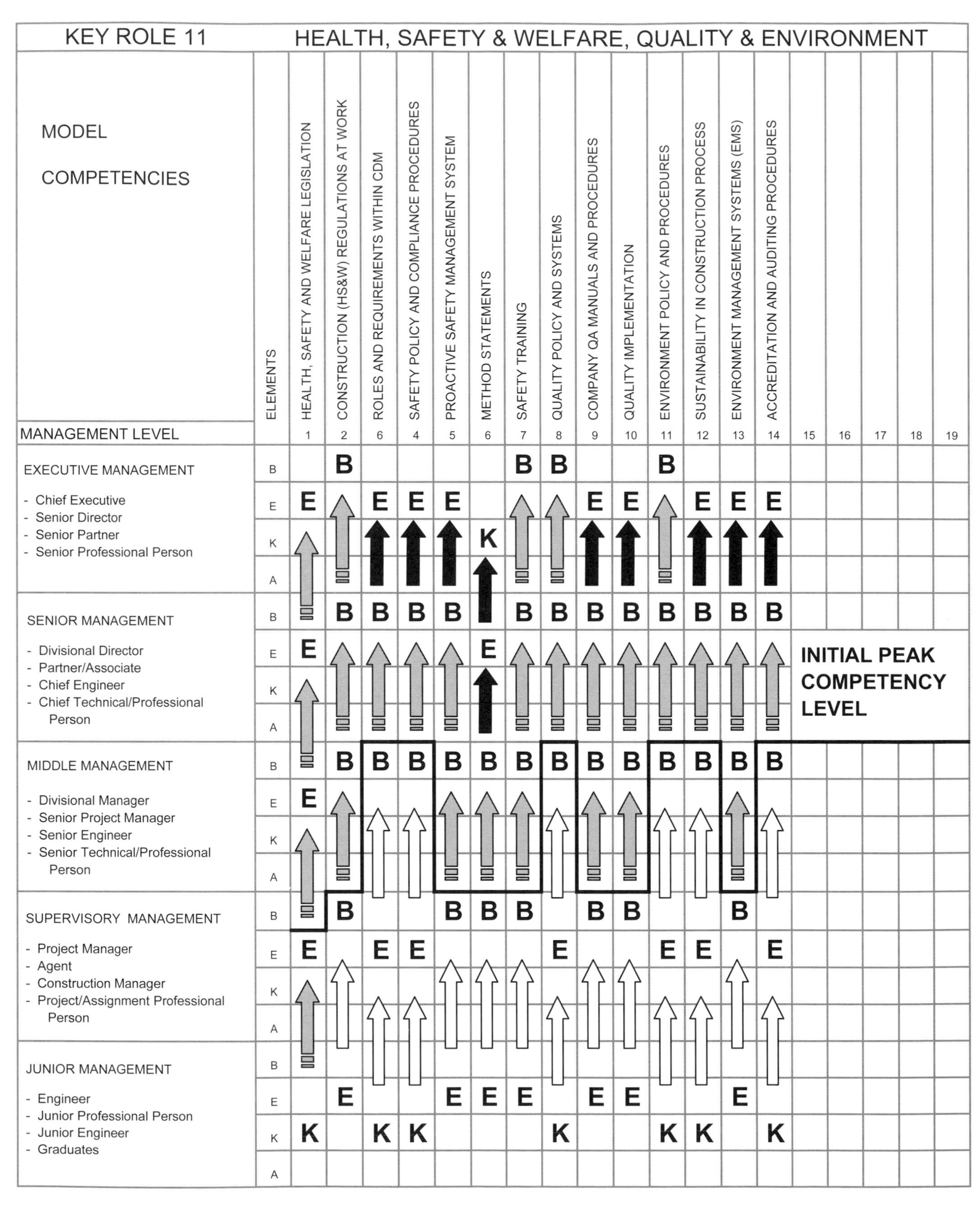

LEVELS OF COMPETENCE

A Appreciation: Know what is meant by the term and what its purpose is.

K Knowledge: Understand in some detail the principles of the topic and how they are applied.

E Experience: Acquisition of knowledge and skill.

B Ability: Application of skill with satisfactory results.

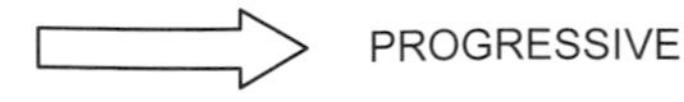 PROGRESSIVE

 SUSTAINED

 REGRESSIVE

MANAGEMENT DEVELOPMENT

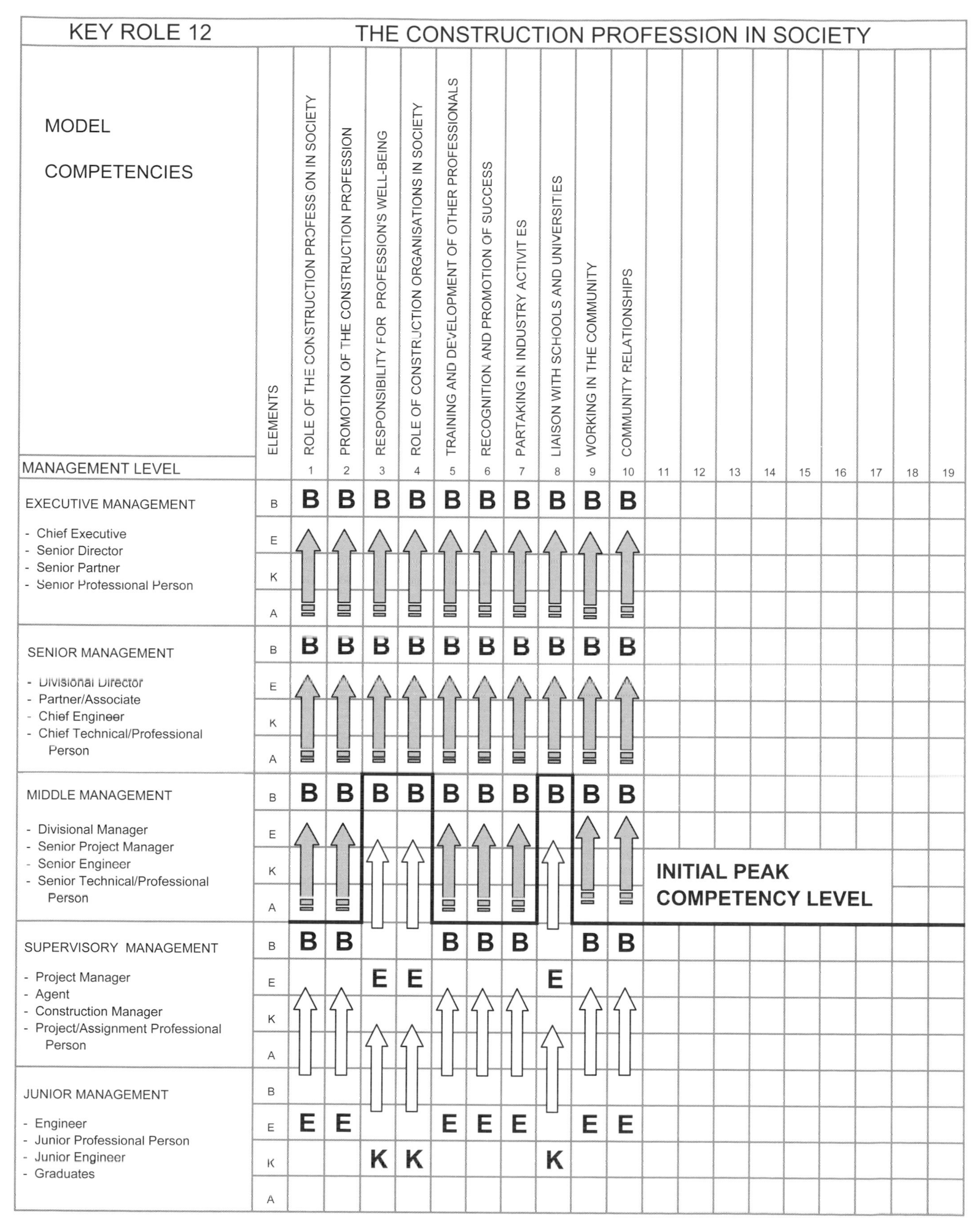

LEVELS OF COMPETENCE

A Appreciation: Know what is meant by the term and what its purpose is.

E Experience: Acquisition of knowledge and skill.

K Knowledge: Understand in some detail the principles of the topic and how they are applied.

B Ability: Application of skill with satisfactory results.

 PROGRESSIVE  SUSTAINED  REGRESSIVE

8. Action plan

To take full advantage of the procedures described in section 6, you must determine a course of action for your further development.

Flow chart

As shown in the flow chart in Fig. 3, your first need is to share the conclusions you reached by self-assessment with someone else.

Discussion

If you are employed you can talk your conclusions over with your line manager, other colleagues, your training department or someone else whose opinion you value. If you have an appraisal procedure this would be an excellent occasion to seek confirmation that your employer agrees with how you see your development.

Fig. 3. An action plan

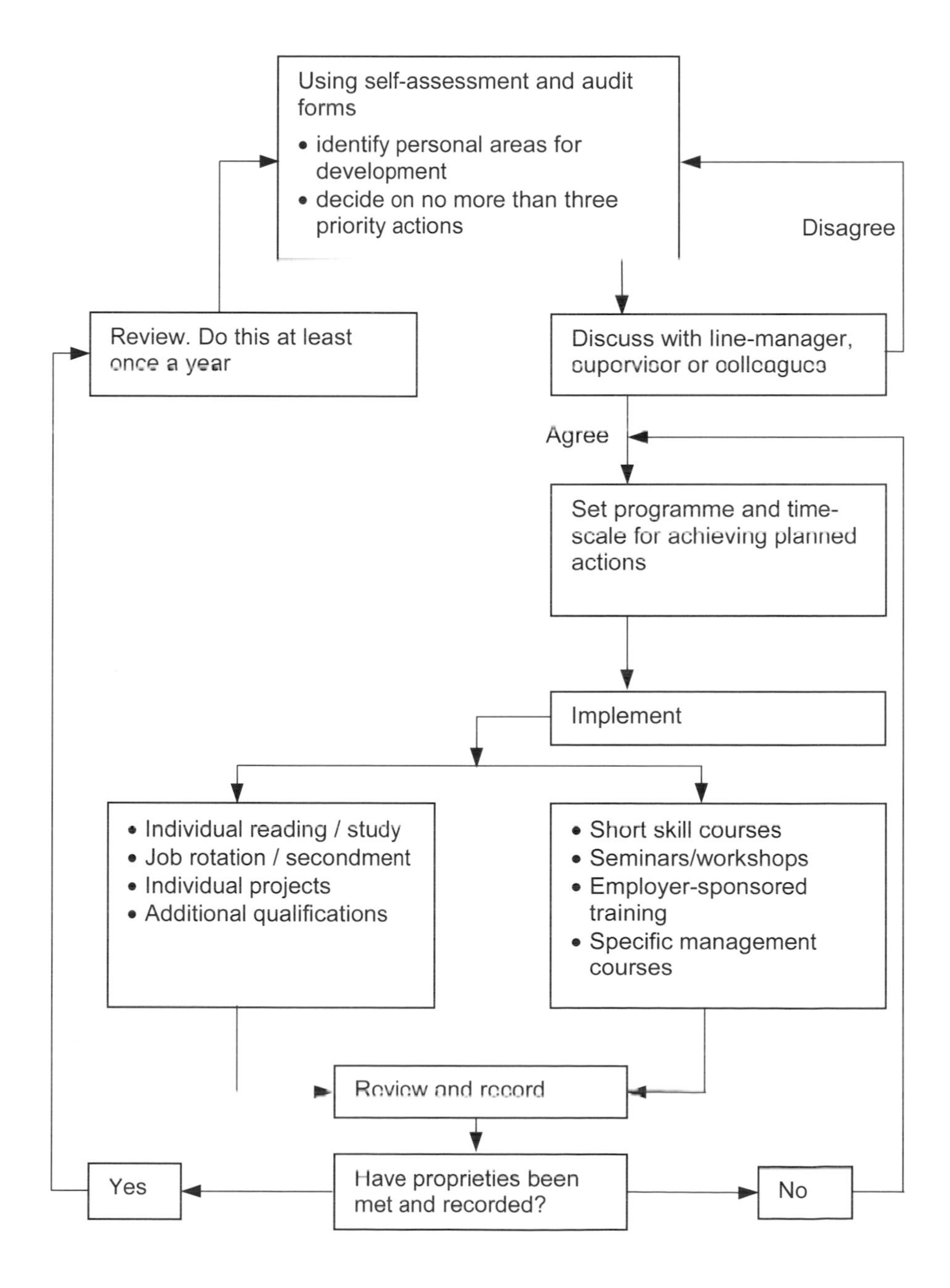

Disagreement

In discussions, the priorities you have determined might not be agreed, because other elements may be identified as more important or you may not be seen to be as competent as you thought. In these circumstances it is necessary to reconsider your self-assessment to see if you should reorder your priorities.

Agreement, programme and time-scale

If your discussions lead to a degree of agreement, you should establish a time-scale for achieving results. This is not only a good discipline to ensure that you will review progress but also, by sharing your intentions with someone such as your line manager, you ascribe importance to the process and will possibly establish access to any help you may require to meet your objectives.

Implementation

The next requirement is to determine how you will achieve your planned development. This depends on your circumstances, the provisions made by your employer, and, more importantly, on your own motivation. Typical examples of ways you may proceed are given in Fig. 3.

Routine review and recording

As you progress you must get into the habit of reviewing and recording your achievements. Professional CPD forms can be used as a career planner or a record of continuing professional development to remind you of your targeted development and how and when this is met.

Meeting priorities

If you can conclude, when you carry out a review on the dates you scheduled, that you have accomplished one or all of your objectives, repeat the process to establish new priorities. Should you decide you need more time, then establish new dates on which you will review progress.

Annual review

Irrespective of whether or not you have carried out interim reviews, you should at least once a year assess yourself to see how you are progressing and reflect on what you need to achieve in the forthcoming year.

Summary

The flow chart in Fig. 3 sets down a procedure that, if followed, will help you to develop an action plan for your continuing development. The process addresses some essential considerations:

- that a plan for twelve months is considered, which should be on a rolling basis and which should cover a three-year horizon
- that, wherever possible, self-assessment needs and the resulting action plan are confirmed by a totally honest appraisal with a superior or more senior person at work or a colleague whose opinions are respected
- that some from of record and subsequent assessment of achievement is kept; Professional CPD forms generally suits this purpose
- that there are a number of ways in which personal development can be achieved: it does not rely wholly on attending formal training courses
- that an action plan is an essential part of management development for professionals, whether it is after four or forty years of a career.

Glossary of terms used in the elements

Key role 1. Corporate management

1. *Corporate strategy:* The translation of 'vision' into future market, product or service positioning of an organisation to maximise prosperity for the company, shareholders and employees.
2. *Business plans and objectives:* A written statement of the objectives of an organisation in pursuit of the corporate strategy and the detailed action plan to ensure achievement of the requisite annual targets
3. *Organisation and management structure*: The internal arrangement of a business that reflects the corporate culture and provides the support necessary for the achievement of the business plan.
4. *Role of officers*: Appointments and responsibilities of those staff in organisations who are charged with liability in law.
5. *Specialist functions and their management:* The use and management of specialist functions that are required for the proper functioning of an organisation in carrying out its core or specialised business.
6. *Memorandum and articles:* A statement required by law for all organisations seeking to limit their liability and register under the Companies Act that sets down how the requirements are to be met.
7. *Mergers and acquisitions:* Options available to organisations seeking to grow other than organically.
8. *Financial arrangements*: Means by which organisations are able to underwrite their borrowing requirements and distribute their assets to achieve their business plans.
9. *Bond and warranty facilities:* The arranged facilities supply all necessary bonds and warranties in order that the business may function properly.
10. *Political and external influences:* The understanding of the potential effect of outside influences on the management and forward workings of an organisation and their proactive management to allow the corporate strategy to be achieved.
11. *Stock market and City:* The necessary dealings, discussions and PR with the 'City' to effectively communicate the forward progression of an organisation to advance company and shareholders' value.
12. *Values and culture.* The understanding and implementation of standards and benchmarks to measure and direct the intellectual development and progress of an organisation to achieve its corporate strategy.

Key role 2: Business management

1. *Business analysis (SWOTs):* Undertaking a detailed study of the main internal and external factors affecting the business within existing markets and developing an action plan to enhance business prosperity so as to ensure achievement of the corporate strategy and business plan.
2. *Key selling points (KSPs):* Identification and promotion of key business attributes of an organisation to maximise its positioning within existing or future markets.
3. *Key performance indicators (KPIs):* The measurement against market-led performance indicators to ascertain the current effectiveness of a

business and the identification and introduction of measures to enhance future performance.

4. *Benchmarking:* The comparison and use of an organisation's performance measured against other market sector leaders to gauge relative positioning and identify and implement improvements to increase competitive advantage.

5. *Innovation:* The positive attitude of companies and individuals towards continuously seeking ways of improving skills, services or goods to deliver a better product to the client, as well as increasing competitive edge.

6. *Value management and engineering:* The process and systems to measure and deliver innovation.

7. *Achieving best value:* On-going liaison with clients to ensure that the product or service being delivered meets or exceeds their overall needs and requirements.

8. *Supply chain management:* Implementation of long-term close working arrangements with other relevant organisations to supply inclusive services or products and to gain maximum competitive advantage and mutual benefit for all parties.

9. *Formation of joint ventures/alliances:* The formal union between organisations in the delivery of products or services that brings together different skills and disciplines under common cause, philosophy and united strategy.

10. *Role and management of winning work:* The coordination of all relevant aspects of an organisation to focus on the requirements of a client in relation to a particular project or service to be procured and compile a bid that will deliver best value with a competitive advantage.

11. *Development and management of teamworking:* The bringing together of all relevant parties and/or disciplines and ensuring maximum working efficiency, contribution and performance by developing common goals, objectives and mutual benefits.

12. *Interdisciplinary skills management:* The understanding of the complex interfaces between different outlook and working practices while harnessing key skills to advance the cause of delivering the overall product or service.

13. *Change management:* Ability and procedures to identify potential changes in any context and operate a system that allows their resolution and incorporation with minimum disruption, without detracting from the achievement of the overall business or project objectives.

14. *Risk management:* Ability and procedures to identify potential risks in any context and operate a system that defines, shares and allocates their management to the party that is best able to deal with them.

15. *Knowledge management:* Systems to capture and distribute information to all relevant persons and parties while maintaining a master database for use and archiving.

16. *Stakeholder relationships:* The bringing together of the parties that depend on the success of a product or service and working together to match desires, wants and mutual benefits.

17. *Training and development policy:* The importance to a business of setting out and undertaking its appraisal, training and development policy so as to enhance the skills and aptitudes of its staff and create a platform to progress the organisation in its future business plans and objectives.

18. *Operating ethics:* The definition, understanding and implementation of an appropriate 'code' by which an organisation responsibly conducts and develops its business.

Key role 3: Financial and management systems

1. *Reporting systems:* Procedures established to ensure that the day-to-day management of the affairs of an organisation are under overall control and that information is available on which further actions can be taken as necessary.

2. *Establishing a budget:* The forecasting of expenditure and income against which controls are developed.

3. *Cost control systems:* Means by which expenditure is identified, accounted, slowed or accelerated to maximum advantage and to maintain the liquidity of an organisation.

4. *Cash flow:* A measure of the cash available to the business.

5. *Profit and loss account:* A retrospective summary of the actual expenditure/income position relative to the budget position for a financial period.

6. *Balance sheet:* A statement produced at a given time showing the assets and liabilities of the business.

7. *VAT and taxation:* Government-dictated deductions from the profits of an organisation, both direct and indirect.

8. *Project and private finance:* The methods by which clients are able to pay for the work they commission and the effects these have on the release of cash in payment for services, goods or work completed.

9. *Whole life costing:* The method of compiling the overall costs for a project that takes into account the long-term operation and maintenance costs.

10. *EU and government grants:* The availability of sources of help and incentives, for funding and acquisition of work, nationally and in the EU.

Key role 4: Promotion and business development

1. *Marketing strategy and planning:* The use of market knowledge to form a coherent strategy and plan to give direction, substance and means to achieving the business and organisational objectives to short/medium/long-term goals within the business plan.

2. *Market research:* The systematic identification of present and future markets and opportunities, the strengths and weaknesses of competitors and the extent of future business potential.

3. *Promotional techniques:* The ways in which an organisation establishes or maintains its corporate identity to attract new and repeat business from clients, e.g. advertising, visits, seminars, functions, sponsorship.

4. *Proposals and presentations:* The written, verbal or electronic techniques used to secure a business opportunity by invitation or as a consequence of targeted marketing activities.

5. *Public relations:* Those activities aimed at creating heightened corporate or product awareness or those exercised in protection/maintenance of an established image.

6. *Assessing key selling points (KSPs):* Promotion of the key attributes of an organisation or business to give competitive advantage within their appropriate market sectors.

7. *Identification of a key client base:* Research to clearly identify the clients that are or will be key 'players' active in the market sectors that have been identified within the marketing strategy and action plan.

8. *Identification of clients' needs:* The understanding of the main 'drivers' that govern the future prosperity of a client and the deeper understanding of the factors and requirements that can be supported or advanced by the services offered by the construction profession or specific business.

9. *Business relationship development:* The promotion of closer relations with key clients by demonstrating a clear understanding of their 'drivers' and needs and a linkage with an organisation's KSPs in order to deliver a better service that satisfies or exceeds their requirements.

10. *Key account management:* The systems and procedures between a client and supplier to focus the delivery of services that better match the client's needs and requirements.

Key role 5: Communications and presentations

1. *Written communication:* The use of hard copies or electronic means to express ideas clearly and precisely.

2. *Oral communication:* The techniques by which ideas are communicated verbally to groups and individuals.

3. *Telephone communication:* The use of the telephone to communicate and transfer ideas, decisions and business matters.

4. *Management of meetings:* The structure and procedures which ensure that time in meetings is planned and effective, that all the participants contribute, that the purpose is achieved and, subsequently, that the conclusions and resultant actions are reported and understood.

5. *Business presentations:* The use of all formats to effectively communicate business objectives in a professional manner that recognise the interests of clients.

6. *Public meetings:* Events arranged to enable issues of possible concern to a wide audience to be debated and at which the specialist input of professionals is communicated.

7. *Dealing with media and VIPs:* The skills with which a professional acts as a spokesperson, giving information that is factual in a courteous way, while remaining conscious of the implications of what is said and written.

8. *Multi-national communication:* The means by which effective communication between multi-cultural people and businesses is professionally and mutually carried out.

9. *Consultations (public etc.):* The ability to be able to explain, listen and 'take-on board' principles and specific actions or methods that are of concern, or have an importance, to other people or parties to enhance the effectiveness of the product or service being delivered.

Key role 6: The client and relationships

1. *Client focus:* Development of relationships with clients to ensure that their key business objectives, 'drivers' and needs are properly identified and understood.
2. *Delivery to meet clients' needs:* A positive attitude towards using knowledge of the clients in the execution of a project to meet or exceed their expectations.
3. *Client satisfaction and measurement:* The procedures necessary to identify the extent to which the delivery of a project is meeting the expectations of a client and the understanding of where improvements can be made.
4. *Predictability of time and cost:* The understanding of the importance of the ability to deliver projects to time and cost and the management systems and procedures necessary to monitor progress and give the client the confidence of certainty.
5. *Partnering/alliancing:* The implementation of teamworking ethics to involve all appropriate parties and expertise related to a specific project or series of projects to ensure that the best whole-life value is delivered to meet the client's needs.
6. *Team integration:* The practical application to utilise and focus a project's interdisciplinary team to allow all parties to effectively input their expertise to meet the specific goals and objectives.

Key role 7: Respect for people

1. *Interview skills:* Questioning and listening techniques to acquire information in order to assess capabilities and use information obtained as appropriate.
2. *Employment conditions:* Statutory or other rules under which people are employed.
3. *Industrial relations:* A framework established to develop employee welfare and business objectives with mutual benefit.
4. *Stress management:* Recognising the consequences of pressure applied to achieve results and exercising awareness, concern and control on behalf of self and others.
5. *Performance appraisal:* A procedure by which the growth and development of individuals and organisations can be monitored and communicated to mutual advantage.
6. *Training and development:* A structured approach towards enhancing the present and future effectiveness of people in their current work, which matches the future needs of the individuals and the organisation.

7. *Leadership*: The initiative, vision and confidence that encourage others to carry out tasks properly under direction.

8. *Negotiation:* Planned discussion and bargaining that consistently allows acceptable agreement to be reached.

9. *Decision making:* Judgement exercised with responsibility once all relevant factors have been taken into account.

10. *Job evaluation:* The determination of skills required to carry out specific tasks or functions.

11. *Delegation:* The act of giving responsibility and authority for a function or task to another while remaining accountable.

12. *Motivation:* Selection of the appropriate technique with which to generate enthusiasm in self and others to produce the best performance of duties.

13. *Time management:* The establishment of priorities which ensure that objectives are met, fully recognising time as a resource.

14. *Understanding managing diversity:* The ability to recognise and control diverse and unplanned situations and ensure that the business objectives and targets are met or exceeded.

Key role 8: Project management

1. *Precontract activities:* The determination of requirements under the contract, the establishment of communication routes, procedures and construction philosophies to be adopted and overall organisation to be employed to meet the objectives and client needs.

2. *Programming work:* Scheduling of activities and critical path analysis such that any interim sectional completion dates and the overall agreed date for completion can be achieved.

3. *Planning and management of resources:* Identification of the skills, goods and services needed to meet the requirements of a contract to ensure their availability as required by the programme, with the organisational structure required to ensure full control.

4. *Staffing and responsibilities:* The identification of those able and available to perform the appropriate tasks and functions required and briefing them in respect of their duties and accountability.

5. *Sub-contractor management:* The identification, appointment and control of others (organisations and/or individuals) employed to meet the obligations of a contract.

6. *Suppliers' conditions and control:* The identification and appointment, under varying contract conditions (UK and/or internationally based), of organisations to supply materials or equipment with all appropriate 'lead-in' and approval periods and any monitoring and expediting necessary to meet the obligations of a contract.

7. *Productivity management and control:* Identification of all appropriate equipment, resources and skills, with rates of work or outputs, needed to meet the requirements of a contract programme and the periodic assessment of actual performance against planned performance.

8. *Progress monitoring and control:* The periodic assessment of actual performance against planned performance to confirm exact programme

position and identify any necessary corrective measures that may be needed.

9. *Cost control and management:* The establishment and monitoring of systems that enable the actual costs being incurred on a contract to be routinely determined and compared with those budgeted, and corrective actions to be implemented.

10. *Certification and payment:* The procedures under which work done or services rendered are valued and remunerated.

11. *Financial control:* The establishment and monitoring of systems that enable the actual costs and 'value' of the contract to be routinely determined and compared with those budgeted.

12. *Team leadership:* Accountability exercised through appropriate delegation and motivation of others carrying out tasks under direction to achieve given or set objectives.

13. *Client relations:* The establishment and maintenance of professional communications, on a regular or as required basis, with the employer or the employer's agents under a contract.

14. *Integrated design & construct management:* The identification and management of the interfaces between the design and buildability of a project in conjunction with the client's requirements and the best value for money.

15. *Implementation of best practice:* The acceptance of the philosophy of seeking, learning and implementing the best practices from within the industry and/or outside it.

16. *Continuous performance improvement (CPI):* The systems and procedures that continuously monitor all aspects of performance and seek to find ways and means of implementing improvement.

17. *Positive and proactive management:* The positive attitude towards looking ahead and countering potential problem areas or situations for the mutual benefit of all parties involved.

18. *Measurement against targets and goals:* The ability to set out and quantify specific objectives and then measure progress to ensure all actions are taken to enable their achievement.

19. *Project review and post appraisal:* The detailed analysis of a completed project in order to learn, and so implement, any improvements to productivity, delivery and outcome that will enhance future performance.

Key role 9: Professional, commercial and contractual practice

1. *Professional ethics and rules of conduct:* Moral, legal and professional rules under which business is conducted.

2. *Contractual arrangements:* The systems and procedures under which contracts are administered in practice.

3. *Contracts and agreements:* The understanding of the general principles of law, and in particular contract law, that govern the pursuit and performance of business.

4. *Non-adversarial forms of contracts and agreements:* The understanding of the principles and culture of collaborative working and the philosophy that guides the working practices.

5. *Commercial negotiation:* The application of planned discussions and bargaining on behalf of an organisation, clients or others to achieve the maximum advantage.
6. *Conditions of contract:* The terms under which skills, goods or services are provided and paid for, often covered by standard forms of contract published by professional bodies or trade associations.
7. *Conditions of engagement:* The terms under which professional expertise is employed.
8. *Variations and changes:* The systems and procedures to identify variations and changes and provide prompt, speedy and proactive resolution.
9. *Avoidance of disputes:* The early proactive negotiations between all relevant parties to agree cause and effect of any changes or variations or any other matters in order to avoid protracted disputes.
10. *Proactive dispute resolution:* The resolve to agree any dispute and knowledge of the use of mediation and/or adjudication to avoid costly arbitration or litigation.
11. *Arbitration and litigation procedures:* The fallback provisions for resolving disputes of any nature, as defined by the form of contract and statute.
12. *Insurance, bonds and warranties:* The conditions and details of all necessary insurance cover, bonds and warranties to meet legal and contractual requirements and any special cases or risk situations as may specifically be required.
13. *Procurement procedures:* The practices, systems and contracts by which goods and services are acquired.

Key role 10: Information and communication technology

1. *Personal ICT skills:* The development of existing skills to be extended to encompass all aspects required within the various sectors or specialisations of the construction industry and the continuous upgrading to take on board future computer and electronic developments so as to ensure ongoing performance improvement.
2. *Specialised software applicable to role:* Knowledge and full practical understanding of the availability, range and scope of software from computer package designers required to support business administration, functioning and control and procedural systems.
3. *Electronic communications:* The speedy and effective use of all electronic forms of communication and information transfer but without detriment to developing personal contacts and relationships.
9. *Use and management of ICT resources:* Understanding the skills and competencies of the resources to effectively manage the hardware, software and systems maintenance in order to allow the business to function properly and allowing time to investigate the next generation of equipment and systems to improve efficiency.
10. *Use and management of e-commerce:* Understanding and effective application of the use of electronic means to procure skills, goods or services without the loss of contractual safeguards.
11. *Use and management of Internet applications:* Understanding and effective application of the use of the Internet to obtain a wider range of relevant information and its increasing use as a business tool.

12. *Development of IT vision:* Recognition of and action to ensure that the advantages so derived from current and future technology are investigated, realised and implemented to constantly improve business performance.

13. *Management of outsourced ICT services:* Recognition and controlled use of expertise in specialist bureaux, normally with performance-based contracts, to enhance the in-house capabilities to meet business needs.

Key role 11: Health, safety & welfare, quality and environment

1. *Health, safety and welfare legislation:* Recognition of the applicability of existing and pending health, safety and welfare law, the liability of all employers and the responsibilities of individuals.

2. *Construction (HS&W) Regulations at work:* Recognition of the applicability of existing and pending health, safety and welfare law as applied to places of work, the liability of all employers and the responsibilities of individuals.

3. *Roles and requirements within CDM legislation:* Recognition of the applicability of existing and pending CDM law, the liabilities, the duties and roles of the employer and individuals necessary to ensure its proper application.

4. *Safety policy and compliance procedures:* The statutory requirement of employers to state their corporate safety policy and the procedures that ensure it is carried out, and the maintenance of a system to continuously monitor its effectiveness and subsequently initiate updating and amending as necessary.

5. *Proactive safety management system:* Enforceable procedures that ensure proper identification of safety risk prior to the carrying out of any work and the compilation of method statements to ensure working attitudes and practices that take all practical steps to reduce the risk of any operation to within acceptable levels.

6. *Method statements:* The knowledge and experience to plan, programme and write down the details of any pending operation so as to minimise risk and in a practical, simple and understandable way.

7. *Safety training:* The necessity to continuously train all staff and operatives to understand all safety requirements and responsibilities and to promote greater skills for all to ensure the safest working practices.

8. *Quality policy and systems:* Recognition of the need for quality to be consciously addressed and set down as a corporate statement in line with recognised standards, supported by systems appropriate to the achievement of the policy.

9. *Company QA manuals and procedures:* Derived as appropriate from the corporate policy statement, establishing the systems, responsibilities and controls that need to apply to deliver a service or product with zero defects.

10. *Quality implementation:* The methods by which the systems are used and the audit techniques to be employed to ensure compliance.

11. *Environment policy and procedures:* Recognition of the need for all environmental aspects of an organisation's operations to be consciously addressed and set down as a corporate statement in line with recognised standards, supported by systems appropriate to the achievement of the policy.

12. *Sustainability in construction process:* The continual search for better ways of carrying out construction operations to minimise the effect on the environment and incorporate 'recycling' and 're-use' to eliminate wastage and thus maximise long-term environment protection and sustainability.

13. *Environment Management Systems (EMS):* The procedures and audit techniques to be used to ensure practical implementation of all working practices necessary to achieve the intent of the corporate statement.

14. *Accreditation and auditing procedures:* Company-wide processes to ensure attainment of recognised standards of operation and a continuous auditing procedure that ensures compliance with or betterment of these standards.

Key role 12: The construction profession in society

1. *Role of the construction profession in society:* The understanding of the responsibilities and duties that the construction profession as a whole must undertake in the betterment of society.

2. *Promotion of the construction profession:* The necessity to extol the virtues of the profession to bring greater understanding to the general public as to its undertakings and achievements and the betterment which the profession brings to society, and thereby raise its profile and status.

3. *Responsibility for the profession's well-being:* The duty on each individual within the profession to give back to the profession a benefit that will ensure its continued well-being

4. *Role of construction organisations in society:* The responsibility of companies and organisations to give to society.

5. *Training and development of other professionals:* The individual requirement to help other professionals to learn and develop and thus promote continuous performance improvement to them and the profession.

6. *Recognition and promotion of success:* The positive attitude towards promoting the industry's successes.

7. *Partaking in industry activities:* The onus on each individual and organisation and their willingness to help and support the various activities undertaken by the profession to ensure its future success and betterment.

8. *Liaison with schools and universities:* The devotion of time and effort to linking with appropriate or local schools and universities and, by giving talks, presentations, organising visits, etc., encourage future recruits by demonstrating the exciting and rewarding potential of the construction industry.

9. *Working in the community:* The re-investment of time, skills and abilities of construction professionals to understand, enhance and progress the community.

10. *Community relationships:* The skill to be able to effectively communicate and develop deeper relationships with the community so as to enrol them and enlist their support in the benefits to social, economic and environmental improvements that the construction profession can and will deliver.

Self-assessment and audit

MANAGEMENT DEVELOPMENT

KEY ROLE 1	CORPORATE MANAGEMENT																		
SELF-ASSESSMENT AND AUDIT	ELEMENTS	CORPORATE STRATEGY	BUSINESS PLANS AND OBJECTIVES	ORGANISATION AND MANAGEMENT STRUCTURE	ROLES OF OFFICERS	SPECIALIST FUNCTIONS AND THEIR MANAGEMENT	MEMORANDUM AND ARTICLES	MERGERS AND ACQUISITIONS	FINANCIAL ARRANGEMENTS	BOND AND WARRANTY FACILITIES	POLITICAL AND EXTERNAL INFLUENCES	STOCK MARKET AND CITY	VALUES AND CULTURE						
MANAGEMENT LEVEL		1	2	3	4	5	6	7	8	9	10	11	12	13	14	15	16	17	18	19
Management level at which	B																			
Assessment carried out:	E																			
....	K																			
Date:	A																			
Assessment as model																				
Assessment greater than model																				
Assessment less than model																				
ELEMENTS FOR ACTION																				
Management level at which	B																			
Assessment carried out:	E																			
....	K																			
Date:	A																			
Assessment as model																				
Assessment greater than model																				
Assessment less than model																				
ELEMENTS FOR ACTION																				
Management level at which	B																			
Assessment carried out:	E																			
....	K																			
Date:	A																			
Assessment as model																				
Assessment greater than model																				
Assessment less than model																				
ELEMENTS FOR ACTION																				

MANAGEMENT DEVELOPMENT

KEY ROLE 2	BUSINESS MANAGEMENT																			
SELF-ASSESSMENT AND AUDIT	ELEMENTS	BUSINESS ANALYSIS (SWOTs)	KEY SELLING POINTS (KSPs)	KEY PERFORMANCE INDICATORS (KPIs)	BENCHMARKING	INNOVATION	VALUE MANAGEMENT AND ENGINEERING	ACHIEVING BEST VALUE	SUPPLY CHAIN MANAGEMENT	FORMATION OF JOINT VENTURES/ALLIANCES	ROLE AND MANAGEMENT OF WINNING WORK	DEVELOPMENT AND MANAGEMENT OF TEAMWORKING	INTERDISCIPLINARY SKILLS MANAGEMENT	CHANGE MANAGEMENT	RISK MANAGEMENT	KNOWLEDGE MANAGEMENT	STAKEHOLDER RELATIONSHIPS	TRAINING AND DEVELOPMENT POLICY	OPERATING ETHICS	
MANAGEMENT LEVEL		1	2	3	4	5	6	7	8	9	10	11	12	13	14	15	16	17	18	19
Management level at which	B																			
Assessment carried out:	F																			
. . . .	K																			
Date:	A																			
Assessment as model																				
Assessment greater than model																				
Assessment less than model																				
ELEMENTS FOR ACTION																				
Management level at which	B																			
Assessment carried out:	E																			
. . . .	K																			
Date:	A																			
Assessment as model																				
Assessment greater than model																				
Assessment less than model																				
ELEMENTS FOR ACTION																				
Management level at which	B																			
Assessment carried out:	E																			
. . . .	K																			
Date:	A																			
Assessment as model																				
Assessment greater than model																				
Assessment less than model																				
ELEMENTS FOR ACTION																				

MANAGEMENT DEVELOPMENT

KEY ROLE 3 — FINANCIAL AND MANAGEMENT SYSTEMS

SELF-ASSESSMENT AND AUDIT	ELEMENTS	REPORTING SYSTEMS	ESTABLISHING A BUDGET	COST CONTROL SYSTEMS	CASH FLOW	PROFIT AND LOSS ACCOUNT	BALANCE SHEETS	VAT AND TAXATION	PROJECT AND PRIVATE FINANCE	WHOLE LIFE COSTING	EU AND GOVERNMENT GRANTS									
MANAGEMENT LEVEL		1	2	3	4	5	6	7	8	9	10	11	12	13	14	15	16	17	18	19
Management level at which	B																			
Assessment carried out:	E																			
....	K																			
Date:	A																			
Assessment as model																				
Assessment greater than model																				
Assessment less than model																				
ELEMENTS FOR ACTION																				
Management level at which	B																			
Assessment carried out:	E																			
....	K																			
Date:	A																			
Assessment as model																				
Assessment greater than model																				
Assessment less than model																				
ELEMENTS FOR ACTION																				
Management level at which	B																			
Assessment carried out:	E																			
....	K																			
Date:	A																			
Assessment as model																				
Assessment greater than model																				
Assessment less than model																				
ELEMENTS FOR ACTION																				

MANAGEMENT DEVELOPMENT

KEY ROLE 4 PROMOTION AND BUSINESS DEVELOPMENT

SELF-ASSESSMENT AND AUDIT MANAGEMENT LEVEL	ELEMENTS	MARKETING STRATEGY AND PLANNING 1	MARKET RESEARCH 2	PROMOTIONAL TECHNIQUES 3	PROPOSALS AND PRESENTATIONS 4	PUBLIC RELATIONS 5	ASSESSING KEY SELLING POINTS (KSPs) 6	IDENTIFICATION OF KEY CLIENT BASE 7	IDENTIFICATION OF CLIENTS' NEEDS 8	BUSINESS RELATIONSHIP DEVELOPMENT 9	KEY ACCOUNT MANAGEMENT 10	11	12	13	14	15	16	17	18	19
Management level at which	B																			
Assessment carried out:	E																			
. . . .	K																			
Date:	A																			
Assessment as model																				
Assessment greater than model																				
Assessment less than model																				
ELEMENTS FOR ACTION																				
Management level at which	B																			
Assessment carried out:	E																			
. . . .	K																			
Date:	A																			
Assessment as model																				
Assessment greater than model																				
Assessment less than model																				
ELEMENTS FOR ACTION																				
Management level at which	B																			
Assessment carried out:	E																			
. . . .	K																			
Date:	A																			
Assessment as model																				
Assessment greater than model																				
Assessment less than model																				
ELEMENTS FOR ACTION																				

MANAGEMENT DEVELOPMENT

KEY ROLE 5	COMMUNICATIONS AND PRESENTATIONS																			
SELF-ASSESSMENT AND AUDIT	ELEMENTS	WRITTEN COMMUNICATION	ORAL COMMUNICATION	TELEPHONE COMMUNICATION	MANAGEMENT OF MEETINGS	BUSINESS PRESENTATIONS	PUBLIC MEETINGS	DEALING WITH MEDIA AND VIPs	MULTI-NATIONAL COMMUNICATION	CONSULTATIONS (PUBLIC ETC.)										
MANAGEMENT LEVEL		1	2	3	4	5	6	7	8	9	10	11	12	13	14	15	16	17	18	19
Management level at which	B																			
Assessment carried out:	E																			
....	K																			
Date:	A																			
Assessment as model																				
Assessment greater than model																				
Assessment less than model																				
ELEMENTS FOR ACTION																				
Management level at which	B																			
Assessment carried out:	E																			
....	K																			
Date:	A																			
Assessment as model																				
Assessment greater than model																				
Assessment less than model																				
ELEMENTS FOR ACTION																				
Management level at which	B																			
Assessment carried out:	E																			
....	K																			
Date:	A																			
Assessment as model																				
Assessment greater than model																				
Assessment less than model																				
ELEMENTS FOR ACTION																				

MANAGEMENT DEVELOPMENT

KEY ROLE 6	THE CLIENT AND RELATIONSHIPS																			
SELF-ASSESSMENT AND AUDIT	ELEMENTS	CLIENT FOCUS	DELIVERY TO MEET CLIENTS' NEEDS	CLIENT SATISFACTION AND MEASUREMENT	PREDICTABILITY OF TIME AND COST	PARTNERING/ALLIANCING	TEAM INTEGRATION													
MANAGEMENT LEVEL		1	2	3	4	5	6	7	8	9	10	11	12	13	14	15	16	17	18	19
Management level at which	B																			
Assessment carried out:	E																			
....	K																			
Date:	A																			
Assessment as model																				
Assessment greater than model																				
Assessment less than model																				
ELEMENTS FOR ACTION																				
Management level at which	B																			
Assessment carried out:	E																			
....	K																			
Date:	A																			
Assessment as model																				
Assessment greater than model																				
Assessment less than model																				
ELEMENTS FOR ACTION																				
Management level at which	B																			
Assessment carried out:	E																			
....	K																			
Date:	A																			
Assessment as model																				
Assessment greater than model																				
Assessment less than model																				
ELEMENTS FOR ACTION																				

MANAGEMENT DEVELOPMENT

KEY ROLE 7 — RESPECT FOR PEOPLE

SELF-ASSESSMENT AND AUDIT	ELEMENTS	INTERVIEW SKILLS	EMPLOYMENT CONDITIONS	INDUSTRIAL RELATIONS	STRESS MANAGEMENT	PERFORMANCE APPRAISALS	TRAINING AND DEVELOPMENT	LEADERSHIP	NEGOTIATION	DECISION MAKING	JOB EVALUATION	DELEGATION	MOTIVATION	TIME MANAGEMENT	UNDERSTANDING MANAGING DIVERSITY					
MANAGEMENT LEVEL		1	2	3	4	5	6	7	8	9	10	11	12	13	14	15	16	17	18	19
Management level at which	B																			
Assessment carried out:	E																			
. . . .	K																			
Date:	A																			
Assessment as model																				
Assessment greater than model																				
Assessment less than model																				
ELEMENTS FOR ACTION																				
Management level at which	B																			
Assessment carried out:	E																			
. . . .	K																			
Date:	A																			
Assessment as model																				
Assessment greater than model																				
Assessment less than model																				
ELEMENTS FOR ACTION																				
Management level at which	B																			
Assessment carried out:	E																			
. . . .	K																			
Date:	A																			
Assessment as model																				
Assessment greater than model																				
Assessment less than model																				
ELEMENTS FOR ACTION																				

MANAGEMENT DEVELOPMENT

KEY ROLE 8 PROJECT MANAGEMENT

SELF-ASSESSMENT AND AUDIT	ELEMENTS	PRECONTRACT ACTIVITIES	PROGRAMMING WORK	PLANNING AND MANAGEMENT OF RESOURCES	STAFFING AND RESPONSIBILITIES	SUB-CONTRACTOR MANAGEMENT	SUPPLIERS' CONDITIONS AND CONTROL	PRODUCTIVITY MANAGEMENT AND CONTROL	PROGRESS MONITORING AND CONTROL	COST CONTROL AND MANAGEMENT	CERTIFICATION AND PAYMENT	FINANCIAL CONTROL	TEAM LEADERSHIP	CLIENT RELATIONS	INTEGRATED DESIGN & CONSTRUCT MANAGEMENT	IMPLEMENTATION OF BEST PRACTICE	CONTINUOUS PERFORMANCE IMPROVEMENT (CPI)	POSITIVE AND PROACTIVE MANAGEMENT	MEASUREMENT AGAINST TARGETS AND GOALS	PROJECT REVIEW AND POST APPRAISAL
MANAGEMENT LEVEL		1	2	3	4	5	6	7	8	9	10	11	12	13	14	15	16	17	18	19
Management level at which	B																			
Assessment carried out:	E																			
....	K																			
Date:	A																			
Assessment as model																				
Assessment greater than model																				
Assessment less than model																				
ELEMENTS FOR ACTION																				
Management level at which	B																			
Assessment carried out:	E																			
....	K																			
Date:	A																			
Assessment as model																				
Assessment greater than model																				
Assessment less than model																				
ELEMENTS FOR ACTION																				
Management level at which	B																			
Assessment carried out:	E																			
....	K																			
Date:	A																			
Assessment as model																				
Assessment greater than model																				
Assessment less than model																				
ELEMENTS FOR ACTION																				

MANAGEMENT DEVELOPMENT

KEY ROLE 9	PROFESSIONAL, COMMERCIAL AND CONTRACTUAL PRACTICE																			
SELF-ASSESSMENT AND AUDIT	ELEMENTS	PROFESSIONAL ETHICS AND RULES OF CONDUCT	CONTRACTUAL ARRANGEMENTS	CONTRACTS AND AGREEMENTS	NON-ADVERSARIAL FORMS OF CONTRACTS/AGREEMENTS	COMMERCIAL NEGOTIATION	CONDITIONS OF CONTRACT	CONDITIONS OF ENGAGEMENT	VARIATIONS AND CHANGES	AVOIDANCE OF DISPUTES	PROACTIVE DISPUTE RESOLUTION	ARBITRATION AND LITIGATION PROCEDURES	INSURANCE, BONDS AND WARRANTIES	PROCUREMENT PROCEDURES						
MANAGEMENT LEVEL		1	2	3	4	5	6	7	8	9	10	11	12	13	14	15	16	17	18	19
Management level at which	B																			
Assessment carried out:	E																			
. . . .	K																			
Date:	A																			
Assessment as model																				
Assessment greater than model																				
Assessment less than model																				
ELEMENTS FOR ACTION																				
Management level at which	B																			
Assessment carried out:	F																			
. . . .	K																			
Date:	A																			
Assessment as model																				
Assessment greater than model																				
Assessment less than model																				
ELEMENTS FOR ACTION																				
Management level at which	B																			
Assessment carried out:	E																			
. . . .	K																			
Date:	A																			
Assessment as model																				
Assessment greater than model																				
Assessment less than model																				
ELEMENTS FOR ACTION																				

MANAGEMENT DEVELOPMENT

KEY ROLE 10 INFORMATION AND COMMUNICATION TECHNOLOGY

SELF-ASSESSMENT AND AUDIT	ELEMENTS	**PERSONAL COMPETENCE**	PERSONAL ICT SKILLS	SPECIALISED SOFTWARE APPLICABLE TO ROLE	ELECTRONIC COMMUNICATIONS						**MANAGEMENT AND APPRECIATION**	USE AND MANAGEMENT OF ICT RESOURCES	USE AND MANAGEMENT OF E-COMMERCE	USE AND MANAGEMENT OF INTERNET APPLICATIONS	DEVELOPMENT OF IT VISION	MANAGEMENT OF OUTSOURCED ICT SERVICES				
MANAGEMENT LEVEL			1	2	3	4	5	6	7	8		9	10	11	12	13	14	15	16	17
Management level at which	B																			
Assessment carried out:	E																			
. . . .	K																			
Date:	A																			
Assessment as model																				
Assessment greater than model																				
Assessment less than model																				
ELEMENTS FOR ACTION																				
Management level at which	B																			
Assessment carried out:	E																			
. . . .	K																			
Date:	A																			
Assessment as model																				
Assessment greater than model																				
Assessment less than model																				
ELEMENTS FOR ACTION																				
Management level at which	B																			
Assessment carried out:	E																			
. . . .	K																			
Date:	A																			
Assessment as model																				
Assessment greater than model																				
Assessment less than model																				
ELEMENTS FOR ACTION																				

MANAGEMENT DEVELOPMENT

KEY ROLE 11 HEALTH, SAFETY & WELFARE, QUALITY AND ENVIRONMENT

SELF-ASSESSMENT AND AUDIT	ELEMENTS	HEALTH, SAFETY AND WELFARE LEGISLATION	CONSTRUCTION (HS&W) REGULATIONS AT WORK	ROLES AND REQUIREMENTS WITHIN CDM	SAFETY POLICY AND COMPLIANCE PROCEDURES	PROACTIVE SAFETY MANAGEMENT SYSTEM	METHOD STATEMENTS	SAFETY TRAINING	QUALITY POLICY AND SYSTEMS	COMPANY QA MANUALS AND PROCEDURES	QUALITY IMPLEMENTATION	ENVIRONMENT POLICY AND PROCEDURES	SUSTAINABILITY IN CONSTRUCTION PROCESS	ENVIRONMENT MANAGEMENT SYSTEMS (EMS)	ACCREDITATION AND AUDITING PROCEDURES					
MANAGEMENT LEVEL		1	2	3	4	5	6	7	8	9	10	11	12	13	14	15	16	17	18	19
Management level at which	B																			
Assessment carried out:	E																			
. . . .	K																			
Date:	A																			
Assessment as model																				
Assessment greater than model																				
Assessment less than model																				
ELEMENTS FOR ACTION																				
Management level at which	B																			
Assessment carried out:	E																			
. . . .	K																			
Date:	A																			
Assessment as model																				
Assessment greater than model																				
Assessment less than model																				
ELEMENTS FOR ACTION																				
Management level at which	B																			
Assessment carried out:	E																			
. . . .	K																			
Date:	A																			
Assessment as model																				
Assessment greater than model																				
Assessment less than model																				
ELEMENTS FOR ACTION																				

MANAGEMENT DEVELOPMENT

KEY ROLE 12 THE CONSTRUCTION PROFESSION IN SOCIETY

SELF-ASSESSMENT AND AUDIT	ELEMENTS	ROLE OF THE CONSTRUCTION PROFESSION IN SOCIETY	PROMOTION OF THE CONSTRUCTION PROFESSION	RESPONSIBILITY FOR PROFESSION'S WELL-BEING	ROLE OF CONSTRUCTION ORGANISATIONS IN SOCIETY	TRAINING AND DEVELOPMENT OF OTHER PROFESSIONALS	RECOGNITION AND PROMOTION OF SUCCESS	PARTAKING IN INDUSTRY ACTIVITIES	LIAISON WITH SCHOOLS AND UNIVERSITIES	WORKING IN THE COMMUNITY	COMMUNITY RELATIONSHIPS									
MANAGEMENT LEVEL		1	2	3	4	5	6	7	8	9	10	11	12	13	14	15	16	17	18	19
Management level at which	B																			
Assessment carried out:	E																			
....	K																			
Date:	A																			
Assessment as model																				
Assessment greater than model																				
Assessment less than model																				
ELEMENTS FOR ACTION																				
Management level at which	B																			
Assessment carried out:	E																			
....	K																			
Date:	A																			
Assessment as model																				
Assessment greater than model																				
Assessment less than model																				
ELEMENTS FOR ACTION																				
Management level at which	B																			
Assessment carried out:	E																			
....	K																			
Date:	A																			
Assessment as model																				
Assessment greater than model																				
Assessment less than model																				
ELEMENTS FOR ACTION